# RELIABILITY AND FAILURE ANALYSIS OF HIGH-POWER LED PACKAGING

WOODHEAD PUBLISHING SERIES IN ELECTRONIC AND OPTICAL MATERIALS

# RELIABILITY AND FAILURE ANALYSIS OF HIGH-POWER LED PACKAGING

**CHER MING TAN**

Center for Reliability Sciences and Technologies, Department of Electronic Engineering, Chang Gung University, Taiwan

Center for Reliability Engineering, Ming Chi University of Technology, Taiwan

**PREETPAL SINGH**

Epistar Corporation, Hsinchu City, East Dist., Taiwan

ELSEVIER

WP
WOODHEAD PUBLISHING
An imprint of Elsevier

Woodhead Publishing is an imprint of Elsevier
50 Hampshire Street, 5th Floor, Cambridge, MA 02139, United States
The Boulevard, Langford Lane, Kidlington, OX5 1GB, United Kingdom



ISBN: 978-0-12-822408-3

For information on all Woodhead Publishing publications visit our website at https://www.elsevier.com/books-and-journals

*Publisher:* Matthew Deans
*Acquisitions Editor:* Stephen Jones
*Editorial Project Manager:* Andrea R. Dulberger
*Production Project Manager:* Prem Kumar Kaliamoorthi
*Cover Designer:* Christian Bilbow



Typeset by TNQ Technologies

# Contents

# Authors' biographies

## Cher Ming Tan

Dr. Tan received his Ph.D. in Electrical Engineering from the University of Toronto in 1992. He has ten years of working experiences in reliability in electronic industry before joining Nanyang Technological University (NTU) as faculty member in 1996 till 2014. He then joined Chang Gung University, Taiwan, and set up a research Center on Reliability Sciences and Technologies in Taiwan and acts as Center Director. He is Professor in Electronic Department of Chang Gung University, Honorary Chair Professor in Ming Chi University of Technology, Taiwan. He has published 400+ International Journal and Conference papers and has given 16 keynote talks and 70 invited talks in International Conferences and several tutorials in International Conferences. He holds 14 patents and one copyright on reliability software. He has written seven books and five book chapters in the field of reliability. He is an Editor of Scientific Report, Nature Publishing Group, an Editor of IEEE TDMR, Series Editor of SpringerBrief in Reliability, Editor of MDPI Applied Research, Editor of Journal of Frontier of thin film, and Associate editor of Microelectronic Reliability. He is a member of the advisory panel of Elsevier Publishing Group. He is also in the committee of IEEE EDS Reliability Physics and the committee of IEEE EDS Master and Ph.D. fellowship selection. He is also named as among the top 2% scientists in the world according to Stanford University research in 2021.

He is a past chair of IEEE Singapore Section, senior member of IEEE and ASQ, Distinguished Lecturer of IEEE Electronic Device Society on reliability, Founding Chair of IEEE Nanotechnology Chapter – Singapore Section, Fellow of Institute of Engineers, Singapore, and Fellow of Singapore Quality Institute. He is the Founding Chair of IEEE International Conference on Nanoelectronics, General Chair of ANQ Congress 2014. He is a recipient of IEEE Region 10 Outstanding Volunteer Award in 2011. He was Guest Editor of International J. of Nanotechnology,

Nano-research letter, and Microelectronic Reliability. He is on the reviewer board of several International Journals such as Thin Solid Film, Microelectronic Reliability, various IEEE Transactions, Reliability Engineering and System Safety etc. He is the first individual recipient of Ishikawa-Kano Quality Award in Singapore in 2014. He is active in providing consultation to multinational corporations on reliability.

His research interests include reliability and failure physics modeling of electronic components and systems, finite element modeling of materials degradation, statistical modeling of engineering systems, nanomaterials and devices reliability, and prognosis and health management of engineering system.

For more detail, please visit www.chermingtan.com.

## Preetpal Singh

Dr. Preetpal Singh received his B.S. degree from the Guru Nanak Engineering College, Hyderabad, India, in 2007, and the M.S. degree from Amity University, Noida, India, in 2013. He finished his Ph.D. degree in 2018 with the Department of Electronic Engineering and Center for Reliability Sciences and Technologies (CReST), Chang Gung University, Taoyuan, Taiwan. He then joined Center for Reliability Engineering, Ming Chi University, Taiwan, and Center for Reliability Sciences and Technologies, Chang Gung University, Taiwan, as a Postdoctoral Researcher from May, 2018 to February, 2021. He has published his research in reputed journals such as the IEEE-TDMR, Microelectronics Reliability, Optical Materials, International Journal of Nanotechnology, ECS Journal of Solid-State Science and Technology, IOP Science Nanotechnology, mdpi Applied Sciences and Scientific Reports. He holds a US patent for his work on High-Power LEDs substrate-GaN thermal and lattice mismatch improvement. He is also the Reviewer for the IEEE-TDMR, IEEE-Access, Scientific Reports, and microelectronics journals. He has been awarded for Best paper award in ICCC ICAI Conference in 2016. He had also held the position of Technical Co-Chair for ninth IEEE International Nanoelectronics Conference in Kuching, Malaysia.

He also collaborated in many industrial projects over the years in solving the complex real-time problems with companies such as Macroblock (Taiwan), Epistar (Taiwan), MOXA (Taiwan), etc.

Presently, he is working with Epistar, Taiwan, as Senior Research Engineer and is responsible for development of different types of LEDs. In 2021, he joined Epistar, Taiwan, as Senior Research Engineer where he is indulged in designing, simulation, fabrication, and failure analysis of various types of LEDs.

His research interests include reliability and failure physics modeling of LEDs, finite element modeling of device degradation, graphene-based high-power LEDs, high-power LED degradation study, and LED reliability.

# Preface

LEDs are recognized as the second revolution in lighting history due to their low cost, low carbon dioxide emissions, and longer lifetime. These benefits over its predecessor lighting options led to increase in application areas and rise in revenue for the LED companies. These new application areas led to the use of LEDs in extreme harsh conditions such as space, mining, glaciers, etc., and some of which requires high reliability of the LEDs. Therefore, traditional techniques to identify failure mechanisms and the statistical tools have to be updated with the changing expectation of LED lifetimes in various emerging application environments. Furthermore, the present-day LEDs are growing in complexity and new materials are introduced to improve certain aspects of LEDs' characteristics such as reflection capability, current flow, heat dissipation, etc. This growing complexity can lead to complex interactions between various LED parts and the occurrence of new failure mechanisms. In other words, the changing world of LEDs' demands requires an updated understanding of failure mechanisms and new failure analysis techniques.

Current LEDs' reliability standards are adopted from the electronic device industry. However, the complex degradation natures of LEDs do not follow the same trends as these electronic devices. Consequently, the prediction of LEDs' lifetime, reliability tests, and their subsequent data analysis must be renewed. A systematic failure analysis is also needed so that failure root causes could be identified and rectified for future applications. In this aspect, finite element analysis has been used in the last two decades for assisting in the determination of the hidden failure causes and the lifetime of the LEDs under extreme environmental techniques.

This book is contributed by researchers who have conducted LEDs' reliability tests and investigated the failure causes for LEDs extensively. This book comprises of the research performed by them wherein new failure techniques are developed and studied. This information is difficult to find in the literature. Providing this knowledge and information in a comprehensive book is good and timely. Most of the content is published in numerous reputed journals or world-class conferences, and they were well cited, which shows the acceptance of the work and high interest within the research community and the present LED industry. Numerous hidden questions about LEDs' failure trends are answered in this book. They introduced new reliability tests, which can cover most stress conditions and implement a systematic failure analysis technique useful in most analysis conditions. Readers of this book will observe the strength of simulation and statistics together in LEDs' reliability studies. The content of this book will help in strengthening the traditional knowledge about LEDs' failure causes and can help LED industry in improving the present LED materials and design according to the applications. It will help LED designers as they need this kind of knowledge from the beginning to avoid failures, which could cause significant profit loss, as the past has shown. From the well-chosen requirement specification to suitable component tests and further up to full system analysis will be made possible with this book. Finally, LEDs are a very good example of optoelectronic devices; hence, even side players may be interested to study and avoid known failures.

We would like to thank the support from Nanyang Technological University, Singapore, and Chang Gung University, Taiwan, for their financial and material support during these researches. We would also like to thank all the coauthors of the manuscripts who helped in one way or another in the compilation of the papers in this book. We would also like to thank the editors and all the reviewers of all the journals such as IEEE Transaction on device and material reliability, Elsevier's Optical Material and Microelectronics reliability, Nature's Scientific reports, etc., who helped in ensuring the high quality of the papers cited in this book. Special mentions include Prof L B Chang of Chang Gung University and Prof H C Kuo from National Chiao Tung University, Taiwan, for their support and discussions during the research.

# CHAPTER 1

# Introduction

## 1.1 Scope of the book

The aim of this book is to let readers understand the evolution of packaging materials and their importance for high-power LED durability and the limitations of current packaging materials. This book also discusses the limitations of the current reliability test standards in determining the lifetime of high-power LEDs owing to insufficient understanding of the packaging materials and their interaction with one another.

Several new failure mechanisms are investigated and presented in different chapters when the LEDs are stressed under different environmental conditions. These failure mechanisms are investigated at both the macro and micro levels. The chemistry underlying the material degradation and its impact on the LEDs are discussed in this book. Understanding of these failure mechanisms at detail levels will help the industry and the researchers in LEDs-related field to develop improved materials for LED packaging. Many failure mechanisms presented in this book are dealt with for the first time, and it will provide a new insight for readers regarding the different possible failure mechanisms in high-power LEDs.

This book also shows the importance of simulation in understanding the failure mechanisms in LEDs. Along with simulation, the use of various destructive and nondestructive tools such as C-SAM, SEM, FTIR, Optical Microscopy, etc., in investigating the causes of LED failures is also presented in this book.

The advancement of LEDs in the last two decades has opened vast new applications for LEDs, which have also led to a harsher stress conditions that high-power LEDs have to face. Thus, existing standards and reliability tests need to be revised to meet the new demands for high-power LEDs. This book serves to provide some insights to researchers and the LED industry, who follow the present standards for LED reliability study on the new failure mechanisms possible for high-power LEDs.

*Reliability and Failure Analysis of High-Power LED Packaging*
ISBN 978-0-12-822408-3
https://doi.org/10.1016/B978-0-12-822408-3.00004-6

Although this book focuses primarily on LED packaging products, the information presented is equally useful and applicable to the packaging of other optical devices such as optical sensors, optical detectors, optical couplers, laser diodes, and so on.

## 1.2 Operation environments of packaged LED

Since their commercialization in the mid-1990s, LED applications have been rapidly expanding [1–3]. This continuous growth in the packaged LED revenues is associated with its many new applications. The efficacy of the packaged white LEDs is now far better when compared to its counterpart fluorescent lighting, and the adoption of LEDs will have a significant impact on our economies, energy consumption, environment, and our living [4–6].

LEDs were adopted for general lighting initially, and the colored and white LEDs then slowly enter into different aspects such as automobile lighting, exterior lighting, backlighting for mobile devices, traffic lighting, lighting in marine and aerospace industry sectors [7,8]. With the many advantages of high-power LEDs including high luminous efficiency, low heat generation, longer lifetime, good mechanical robustness, low voltage operations, and eco-compatibility [9], they are no longer confined to indoor lighting and application areas of LEDs are now become wider. LED market is growing rapidly as shown in Fig. 1.1, and its biggest share is lighting followed by traffic lights currently. As shown in Fig. 1.1, the LED market is rapidly expanding, with lighting taking the lead, followed by traffic lights. LEDs have seen a rapid increase in outdoor applications over the years, while indoor applications have been decreasing. In outdoor applications, LEDs are now exposed to harsher environmental conditions, such as higher temperatures, higher humidity levels, and cyclic pressures, which were not present when they were first introduced.

Therefore, LEDs and their packaging materials face significant challenges as a result of these stress situations, and the aim of this chapter is to review these challenges and provide recommendations for future improvements.

## 1.3 LED packaging challenges

In the past, high-power LEDs had low efficacy. With technology advancement, their efficacy improves, and high-power LEDs' application

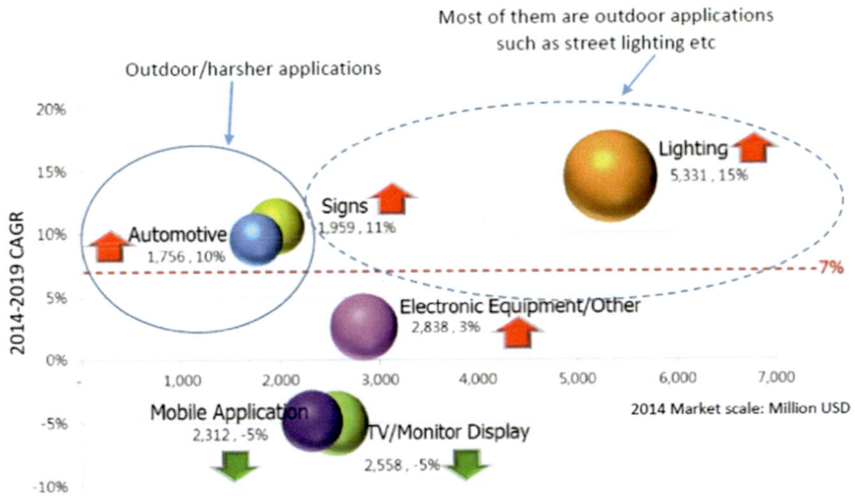


**Figure 1.1** High-brightness LEDs market size distribution and segmentation [10]. *(Microelectronics Reliability, Elsevier. Reprinted with permission.)*

areas are expanding, but this poses various challenges for their packaging materials to ensure their durability.

Most high-power LEDs, in particular, can only convert 15%–20% of input electrical power into light [11], with the rest being converted to heat. This renders a rise in LED junction temperatures, making thermal management a major packaging challenge [12,13]. Advanced methods for enhancing heat removal and minimizing material coefficient of thermal expansivity (CTE) mismatch are important topics to address for high-power LED packaging [14,15].

As shown in Fig. 1.2 [16], a packaged high-power LED contains many components, including the molding part, heat sink, LED chip, die attach, lead frame, encapsulant, Au wires for bonding, and lens.

Most LEDs used epoxy as an encapsulant during the early stages of LED's growth, which degrades quickly when exposed to high temperatures or UV light, opening pathways for moisture and air infiltration [17,18]. Epoxy-based systems also tend to exhibit discoloration during their lifetime, which reduces the overall light output of LEDs. These degradation processes render short lifetime of high-power LEDs [19,20].

Due to their broad range of advantages, such as a suitable refractive index of 1.4–1.57, good UV and thermal resistance, low impurity levels, good moisture resistivity, high optical clarity, and low modulus, silicone-based materials are considered to be the most acceptable materials for

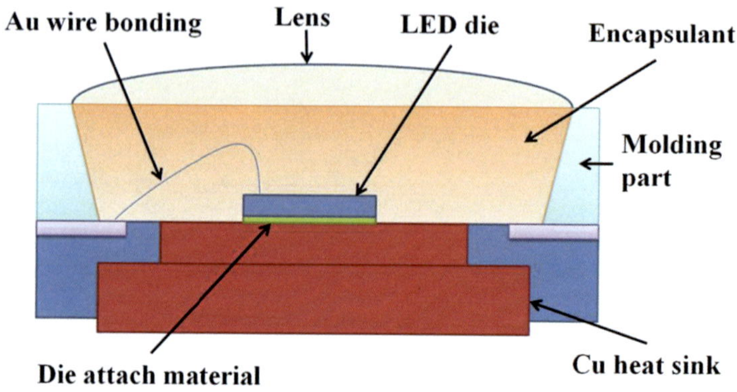


**Figure 1.2** A cross-sectional view of an SMD LED package in schematic form [16]. *(IEEE Access, IEEE. Reprinted with permission.)*

encapsulant, molding component, and lens of LEDs [21,22]. Silicone is used as a molding component for the housing material in LEDs because of its mechanical strength and light performance enhancement due to its reflector shape and unique filler addition [23].

In the LED industry, silicone-based packaging has replaced thermoplastic-based premolded packaging since thermoplastic-based molding packaging is more prone to surface yellowing (browning) and encapsulant detachment under high temperature exposure, resulting in shorter LED lifetime [24,25].

Besides thermal management and light extraction efficiency, moisture is also an important factor, which should be considered along with high temperature as their combination can lead to rapid lumen degradation due to several failure mechanisms as studied by Tan et al. [26]. Tan et al. [23], for example, discovered that moisture penetrated the silicone encapsulant and caused rapid lumen degradation for high-power LEDs. Unfortunately, the number of reported research works on humidity-related LEDs degradation is much less as compared to that on thermal-related studies as shown in Fig. 1.3.

From the above discussion, it can be inferred that following properties of packaging materials are needed to improve the lifetime of LEDs:

1. Encapsulant materials with appropriate refractive index so that most of light can be transmitted out from the encapsulated chips;
2. Packaging materials with high thermal and UV resistance;
3. Packaging materials with a good CTE match with their surrounding materials and the LED chip; and

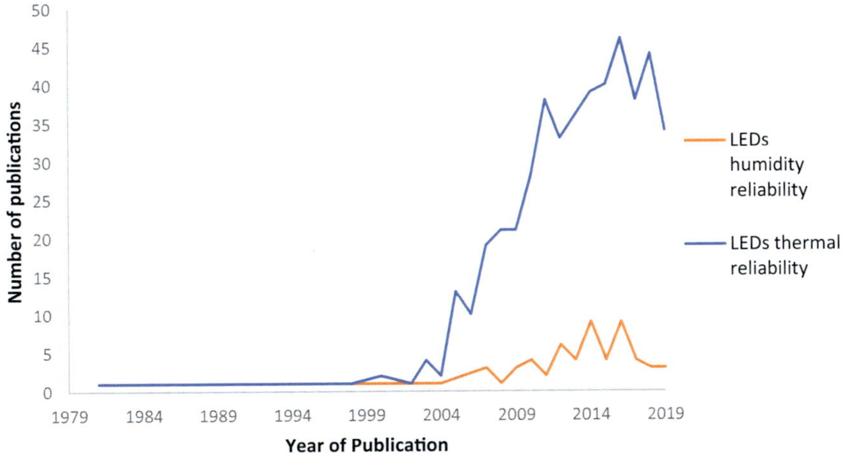


**Figure 1.3** Comparison of number of publications on LED's humidity and thermal reliability studies respectively over the years. Data is collected from Scopus [23]. *(IEEE 23rd International Symposium on the Physical and Failure Analysis of Integrated Circuits (IPFA). IEEE. Reprinted with permission.)*

4. High-adhesion and moisture-resistant encapsulant and molding component products.

Let's take a look at the history of high-power LED packaging materials for tackling the various issues encountered.

## 1.4 Commonly used packaging materials and their evolution

There are various LED parts that come in the category of packaging materials such as heat sink, molding part, encapsulant, and lead frame for electrical connections. The present book deals mainly with the reliability degradation of LEDs due to molding part and encapsulant as these package materials are still under maturing phase when compared with other LED parts. In this section, we will discuss the various molding and encapsulant materials used in LED industry and their merits and demerits.

### 1.4.1 Encapsulant materials

High-power LEDs operate at currents of 350 A or higher, and in this current range, high amount of heat is generated inside the package. Both the LED chip and the phosphor produce heat within the LEDs package (in the case of white LEDs that use phosphor for light conversion). Therefore, high-power LED packages generally include heat sink and thermally enhanced metal core PCBs as substrate for efficient heat dissipation from LED packages.

In the packaging industry, epoxies are the most commonly used polymer material [27]. Epoxy has special properties such as good optical, electrical, and mechanical properties, as well as corrosion resistance, excellent chemical resistance, low shrinkage, and good adhesion, making it a cost-effective material for LED encapsulation [28]. Epoxy encapsulants have long been the best alternative for low-power LEDs, and they do not degrade over time for LEDs of different wavelengths.

Epoxies are divided into two categories based on their physical forms: liquid and solid (also known as molding compound). Generally, these two forms are used in various packaging methods. As shown in Fig. 1.4 [29], epoxy is a thermoset substance with a three-dimensionally connected polymer network that results from the curing of an epoxy resin with an anhydride hardener. Each epoxide group is consumed by an anhydride hardener moiety during the polyaddition cure reaction stage, which is primarily a thermal process, and forms a covalent bond, resulting in a rigid polyester structure without the harmful by-product being released [30]. In organic chemistry, a moiety is a part of a molecule that is typically given a name as it can be found within other kinds of molecules as well.

Epoxy products possess a low permeability to gases and humidity, making them good candidates for avoiding lead frame corrosion. Epoxies also attain greater hardness when compared to its competitors such as silicone and polycarbonates. However, its lack of thermal and photo stability are the biggest drawbacks for epoxy materials when used in high-power LED packages as epoxies are prone to thermal oxidation [30].

This brings us to Silicone, the second generation of polymer materials used in LED packaging. The chemical structure of polymerized siloxanes or polysiloxanes is $[R2SiO]_n$, in which an inorganic Si—O backbone is bonded with either methyl or phenyl groups [31]. As the Si molecule in the backbone is replaced with C, the total bond strength increases, and silicones have higher thermal and photo stability than epoxies. Silicones have

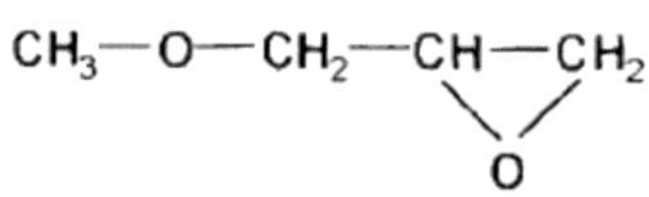


**Figure 1.4** Chemical backbone structure of epoxy [29]. *(Biomaterials, Elsevier. Reprinted with permission.)*

different properties such as refractive index, humidity/gas permeability, and hardness depending on the length of the Si—O chain, the form of side groups attached, and the cross-link ratio. Two types of silicone groups are shown in Fig. 1.5, which are commonly used in LED manufacturing. To achieve specific light output, one or a combination of these two classes of silicones can be used to better fit the necessary properties such as refractive indexes.

Silicones have higher photo and thermal stability than epoxies, making them a good candidate for blue or white high-power LED packages. These materials can undergo degradation under harsh environments especially at high temperature and under humid conditions that can lead to light output degradation in LEDs [16]. Irreversible thermal oxidation process can take place due to methyl group oxidation and results in cross-linking between silicone backbone chains. These processes can result in silicone hardening and cracks as will be discussed later in detail [30]. It is found that LEDs packages with cracks in their packaging will result in changes in the overall light output and higher corrosive gases and moisture penetration [32]. This can in turn lead to further lead frame and wire bonding corrosion-related failures. Apart from this, phenyl-based chromophore formation may also cause silicone discoloration and induce color shift [30]. Thus, it is important

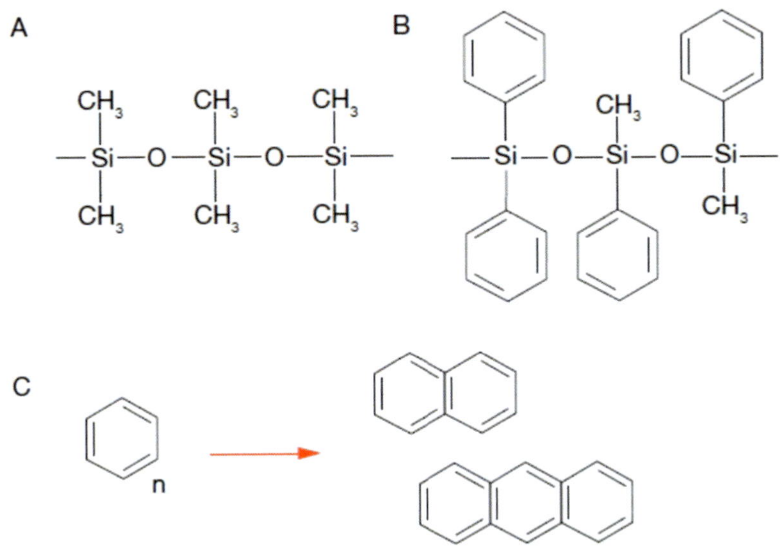


**Figure 1.5** Two silicone examples: Dimethylsiloxanes (A). phenylsiloxanes (B) [16]. *(IEEE Access, IEEE. Reprinted with permission.)*

to choose appropriate silicone material and proper thermal management to reduce such irreversible reactions and the chances of crack formation and color shift. Table 1.1 summarize the advantages and disadvantages of the two most widely used packaging materials for LEDs.

## 1.5 Reliability requirements of packaged LED

Though LED lighting products have advantages over their predecessors in terms of long lifetimes for general lighting and robustness for harsh environment applications, developing generally accepted LED lighting product-level reliability standards is difficult. LED source lumen maintenance test (ANSI/IES LM-80), driver robustness test (ANSI C82.16), catastrophic failure rate prediction (IES TM-26), and lumen maintenance projection have all been the subject of interest in recent years (IES TM-21). However, work related to reliability tests for solid-state lighting (SSL) at the lamp or luminaire stage is yet to be completed.

**Table 1.1** Brief summary of epoxy and silicone properties [32]

| Polymeric material | Advantages | Disadvantages |
|---|---|---|
| Epoxies | • good chemical and mechanical protection<br>• low moisture absorption<br>• suitable for all thermosetting processing methods<br>• excellent wetting characteristics<br>• ability to cure at atmospheric pressure<br>• excellent adhesion to a wide variety of substrates under many environmental conditions<br>• thermal stability up to 200°C | • high stress<br>• moisture sensitivity<br>• short shelf life (can be extended under low-temperature storage conditions) |
| Silicones | • low stresses<br>• excellent electrical properties<br>• good chemical resistance<br>• low water absorption<br>• thermal stability up to 315°C<br>• good UV resistance | • low tensile tear strength<br>• high cost<br>• attacked by halogenated solvents<br>• poor adhesion<br>• long cure time |

Materials and Processes for Electronic Applications, Encapsulation Technologies for Electronic Applications (Second Edition), William Andrew Publishing. Reprinted with permission.

A long debate persists in the LED industry worldwide regarding which standard should be chosen, how long the test be running or how should the test data be interpreted so that the lifetime of the LEDs can be projected? All these are due to the unavailability of common acceptable tests and methods in the industry. There are no common standards for LEDs to be tested under conditions that replicate the real-time scenarios in various applications.

Two common arguments in developing the reliability test standards are as follows. One is "what" should be tested and other is "who" should write the reliability test standards. Due to different environmental factors such as humidity, dust, temperature, chemical, corrosion, vibration, etc., as well as varying application impacts, minimum set of tests or a common denominator has not been agreed upon by different parties. Illuminating Engineering Society of North America (IES) develops test standards and LED-related measurement standards, but concerns are raised whether all lighting products should fit within IES scope or not, and also whether LED products are being overtested if more standards are written. Another concern is regarding the feasibility and validity of tests.

Basically, there are four standards that are adopted and implemented for LEDs' reliability testing globally. These four standards are:

- *International Electrotechnical Commission (IEC) standards*
- *Illumination Engineering Society (IES) standards*
- *ASSIST project standards*
- *JEDEC test standards*

These abovementioned reliability standards help the researchers and companies to perform LED reliability tests and gain knowledge of unknown failure mechanisms and also help in estimating the useful lifetime of the LEDs. More in-depth knowledge of these test standards will be provided in Chapter 9.

## 1.6 Overview of LED failures

Few authors [3] recorded the change in lumen degradation of high-power LEDs over time, and they found that the lumen degradation over time can be divided into three different stages when observed under real-time outdoor conditions, i.e., high temperature and humidity conditions, as follows.

**a.** Initial rapid degradation [3].
**b.** Lumen recovery following initial rapid degradation [3].
**c.** Subsequent permanent degradation [3].

Tan et al. [33] discovered that the initial rapid lumen degradation was caused by moisture penetrating the silicone encapsulant in high-power LEDs. Moisture clusters in the encapsulant diffract the light from the LED chip, decreasing the total amount of light that comes out of the LED encapsulant.

Tan et al. [33] discovered that when the moisture pore size is greater than 50 nm, the percentage attenuation increases significantly. The percentage attenuation is negligible when the pore size is less than 30 nm. To avoid the initial sharp drop in light output in a humid setting, the pore size of the silicone gel should be less than 30 nm.

There was a recovery after the initial rapid lumen degradation. Preetpal and Tan [4] built a finite element model to investigate this recovery, and they discovered that the average moisture content of the entire package rises over time, and that about 100 h (85%RH/85°C), moisture began to penetrate into the die attach, resulting in a decrease in silicone's moisture content as mild delamination at the die attachment acts as a capillary tube, sucking the moisture from the silicone into itself, and the initial attenuation of light output due to moisture in silicone is reduced. Since die attach has a higher moisture absorption capacity than silicone encapsulant, moisture entering the LED package is much less than moisture sucked by the die attach. This provides a way for more light to exit the LED encapsulant without much scattering, resulting in the recovery of LED lumen degradation over time. The simulation findings were confirmed experimentally using the confocal scanning acoustic microscopy (C-SAM) technique, which revealed serious die attach delamination after 100 h of testing [4]. As a result, the LED seems to be recovering, but it was actually degrading internally, as observed.

Following the lumen recovery, the final stage of lumen decrease was observed, with no subsequent recovery. As shown in Fig. 1.6, the degradation process started after the lumen recovery, with bond pad contamination and Kirkendall void formation, galvanic dissolution, phosphor dissolution, and die attachment delamination. The breakdown of the Au wire bond causes a shift in resistance, which leads to a reduction in light intensity [3].

During an accelerated humidity test, the amount of moisture trapped in the die attach/thermal interface material or between the electrical contact pads on the die increased. As heat is produced by the LED dice during the powering of the LED for optical testing, the trapped moisture will be vaporized. According to Tan et al. [23], such vaporization would cause

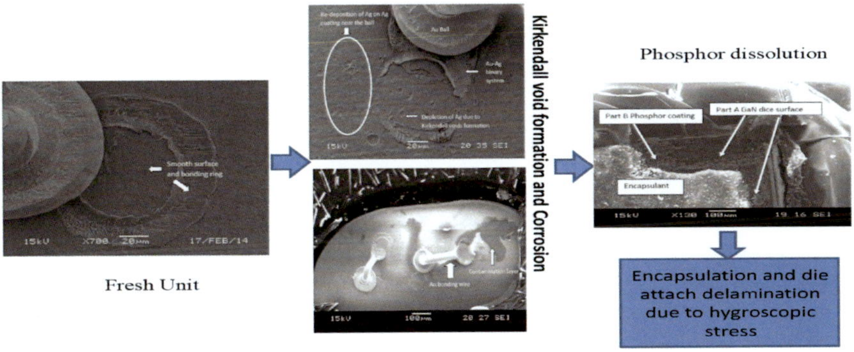


**Figure 1.6** Final degradation processes in high-power LEDs, which are permanent [34]. *(IEEE Transactions on Device and Materials Reliability, IEEE. Reprinted with permission.)*

mechanical damage to the LED dice if a micro-crack was initially present before the time of vaporization. When the LED is turned on at a high ambient temperature and moisture is present at the lens—encapsulant interface, lens cracking can also occurs [35].

While the moisture effect on LED is significant as reported, there is no reliability test to observe the combined effect of temperature, moisture, and electrical stresses on high power LEDs. Hence, Preetpal and Tan [34] developed a new reliability test called MET test to observe the combined effect of temperature, moisture, and electrical stresses on high-power LEDs. There are several failure mechanisms that are peculiar to the MET test, which are not present in the temperature—humidity, temperature alone, and electrical alone tests, as shown in the comparison Table below, and these mechanisms should be studied for LEDs in outdoor applications [23] (Table 1.2).

## 1.7 Statistics used in failure analysis

Each luminaire designer must be aware of the LED's expected long-term performance under the conditions present in their design. Since LEDs have so many different operating conditions, it's difficult for LED manufacturers to monitor all of them. As a result, LED reliability models are required in order to extrapolate their lifetime from a simple collection of data, and extrapolation takes place in two dimensions: operating conditions (such as drive current and board/junction temperature) and time.

**Table 1.2** Comparison of failure mechanisms related to temperature–humidity, temperature alone, electrical alone, and met test [23].

| External stress(es) applied | Failure sites | Failure modes | Failure mechanisms |
|---|---|---|---|
| Electrical stress | Chip level | Lumen degradation, increase in reverse leakage current, increase in parasitic series resistance and short circuit. | Defects in LED chip, electromigration of the metal atoms in the electrical contact to the surface of the LED die, instability in Mg diffusion in p- GaN layer and dislocation generation and movement at chip level |
| | Package level | Lumen degradation | Carbonization of encapsulation and phosphor thermal quenching |
| Thermal stress | Package level | Lumen degradation, Detachment of encapsulant from molding part | Browning of the white reflector molding part of device. |
| | Chip level | Lumen degradation | Degradation of the phosphor's conversion efficiency |
| Electrical stress and thermal stress | Chip level | Lumen depreciation | Crack in the LED chip |
| | Package level | Lumen degradation | Encapsulant yellowing, lens cracking or solder joint fatigue |
| Thermal stress and moisture stress | Chip level | Lumen degradation, Forward voltage change, permanent destruction of LED i.e., no light | Wire ball bond fatigue |
| | Package level | Reduction of the overall light output, | LED chip-die attach delamination, moisture entrapment in encapsulant, and lens cracking |
| Thermal, electrical and moisture | Chip level | Rapid lumen depreciation, forward voltage drop | LED chip/die attach delamination |

**Table 1.2** Comparison of failure mechanisms related to temperature–humidity, temperature alone, electrical alone, and met test [23].—cont'd

| External stress(es) applied | Failure sites | Failure modes | Failure mechanisms |
|---|---|---|---|
| stresses (MET test) | Package level | Rapid lumen degradation | Browning of the white silicone reflector molding part of device, encapsulant detachment from molding part, resulting in large passage of moisture into package |

IEEE 23rd International Symposium on the Physical and Failure Analysis of Integrated Circuits (IPFA), IEEE. Reprinted with permission.

### 1.7.1 Lumen maintenance model

Companies test LEDs under particular conditions to allow accurate extrapolation of lumen maintenance for any operating conditions. The LM-80-08 standard for LED product testing, as established by independent industry body IESNA, specifies three of these requirements [36].

An LED's lumen maintenance must be extrapolated into the future. The LM-80-08 standard calls for a minimum of 6000 h of testing [37]. A minimum of 25 samples is needed by the ENERGY STAR Manufacturer's Guide (September 2009) [37]. Lumileds, for example, bases its lumen maintenance model on data from experiments that lasted much longer and had a much greater sample size than the LM-80-08 and ENERGY STAR requirements. Lumileds also extrapolates lumen maintenance behavior using the same "exponential extrapolation" model that ENERGY STAR uses to forecast 6000-hour limit points.

Using test data collected at higher/accelerated test conditions, this extrapolation model can be used to estimate LED failure rates and time-to-failure at actual operating conditions. To measure time-to-failure, the Arrhenius equation is commonly used for exponential luminous decay. On the other hand, some LED manufacturers employ proprietary models that exaggerate their products' capabilities. The long-term lumen maintenance of a LUXEON Rebel, for example, can be graphed.

Although information about the spread in lumen maintenance behavior across a population of LEDs is not required by LM-80-08, Lumileds can obtain this information from the tests mentioned above. This data can be expressed as a likelihood that the lumen maintenance of a single LUXEON Rebel LED will exceed a certain threshold after a certain amount of time and under certain operating conditions.

### 1.7.2 Catastrophic failure model

Although industry practice focused on the development of lumen maintenance ratings for power LEDs, such as the LM-80-08 standard, there is no such standard protocol for catastrophic failure testing for LEDs.

Lumileds developed a testing protocol that can be used by any LED manufacturer and is currently in use for LUXEON Rebel LEDs. The test takes the cells from the lumen maintenance-testing protocol and applies highly stressed conditions to them. These include reaching the device's maximum temperature of 150°C and current rating of 1A. The aim is to accelerate catastrophic failures in order to develop a reliable predictive catastrophic failure model.

This test set contains cells that operate above the "recommended' and "maximum' levels. The data from these extremely stressed environments contributes to a higher level of trust in extrapolations of LED output at normal operating levels. The operating conditions in cells "Y" are higher than the prescribed operating conditions. The operating conditions in cells "X" are higher than the maximum scores.

Models of a given phenomenon can be constructed by leveraging available data from industrial sources and laboratory experiments. In some instances, the industry has already developed standard methods, such as TM-21-11 [38], for projecting long-term performance based on data collected using standard procedures, such as LM-80-08. However, in many cases, there were no acceptable models or methods to create models, and different approaches were adopted. In general, several approaches such as Empirical, Kaplan–Meier Models, Weibull Models, Physics of Failure, Simulation models, Principle Component analysis, Kalman Filtering Models, etc. All the models have their own advantages and disadvantages. More detailed description along with the examples where these models are employed in case of LED studies is presented in Refs. [36,39].

All the models discussed in this section are used to determine the time-to-failure for LEDs. However, LEDs could fail due to different mechanisms and hence different statistical distributions will have to be employed for these different mechanisms. Singh et al. [40] provided a statistical method for determining the number of potential degradation/failure mechanisms in a given collection of measured data, as well as the likelihood of each unit belonging to each degradation/failure mechanism. They used the Akaike information criterion (AIC) to determine the number of possible degradation mechanisms in a given collection of calculated data, as well as simulated annealing (SA) and the expectation–maximization (EM) algorithm to determine the probability of each unit belonging to each possible degradation/failure mechanism. According to Tan et al. [34], the global maximization algorithm is often used to evaluate the distribution parameters of the statistical distribution associated with each potential degradation/failure process. They discovered that the degradation mechanisms of LEDs at 95%, 85%, and 70% RH under temperature–humidity tests are entirely different, and that using the typical 85% RH at 85C data for lifetime extrapolation, as in the case of integrated circuits where the Peck model is widely used, is unacceptable. The degradation mechanisms of LEDs in high temperature–humidity tests vary significantly from those of integrated circuits. As a result, a new humidity model is urgently needed to estimate the lifetime of high-power LEDs under humid operating conditions. This makes extrapolating test results to standard operating conditions for lifetime prediction difficult. In the following chapters, more details and findings related to this model will be explored in depth.

## 1.8 Summary

This chapter is intended for readers to develop a strong understanding of LEDs packaging materials and its reliability issues. This chapter deals with the expanding market of the LEDs and these increasing applications in turn lead to numerous challenges posed for LEDs packaging industry. Commonly employed packaging materials such as epoxies and silicone are discussed in detail in this chapter including their applications, advantages, disadvantages, and various failure modes under different stress conditions.

This chapter also covers the reliability requirements for packaged LEDs due to various failure mechanisms of LEDs, under various stresses. LEDs' failure statistics is discussed, especially the reliability models generally employed in industry and their implication on the LEDs reliability. It is revealed in this chapter that the commonly used models for the lifetime estimation by extrapolation for integrated circuits are not suitable for LEDs and there is a need for an improved model for LEDs lifetime estimation. In short, this chapter creates a good base for the future journey into the LEDs packaging reliability.

## References


[1] Y. Narukawa, White-light LEDS, Optics & Photonics News 15 (4) (2004) 24–29.

[2] M.G. Craford, LEDs for solid state lighting and other emerging applications: status, trends, and challenges, in: Fifth International Conference on Solid State Lighting, vol 5941, International Society for Optics and Photonics, 2005.

[3] G. Craford, Current State of the art in high brightness LEDs, APS March Meeting Abstracts (2007) J3–1.

[4] F.M. Steranka, et al., High power LEDs–Technology status and market applications, Physica Status Solidi (A) 194 (2) (2002) 380–388.

[5] J.Y. Tsao, et al., Solid-state lighting: an integrated human factors, technology, and economic perspective, Proceedings of the IEEE 98 (7) (2010) 1162–1179.

[6] H. Khorasanizadeh, et al., Energy and economic benefits of LED adoption in Malaysia, Renewable and Sustainable Energy Reviews 49 (2015) 629–637.

[7] M.-H. Chang, et al., Light emitting diodes reliability review, Microelectronics Reliability 52 (5) (2012) 762–782.

[8] S.W. Sanderson, et al., Lighting industry: structure and technology in the transition to solid state, in: 2008 Industry Studies Conference Paper, 2008.

[9] K. Eichhorn, LEDs in automotive lighting, in: Light-Emitting Diodes: Research, Manufacturing, and Applications X, vol 6134, International Society for Optics and Photonics, 2006.

[10] P. Singh, C.M. Tan, A review on the humidity reliability of high power white light LEDs, Microelectronics Reliability 61 (2016) 129–139.

[11] Y.-C. Lin, et al., Materials challenges and solutions for the packaging of high power LEDs, in: 2006 International Microsystems, Package, Assembly Conference Taiwan, IEEE, 2006.

[12] Y. Huaiyu, et al., A review of passive thermal management of LED module, Journal of Semiconductors 32 (1) (2011) 014008.

[13] A. Poppe, C.J.M. Lasance, On the standardization of thermal characterization of LEDs, in: 2009 25th Annual IEEE Semiconductor Thermal Measurement and Management Symposium, IEEE, 2009.

[14] U. Lafont, H. van Zeijl, S. van der Zwaag, Increasing the reliability of solid state lighting systems via self-healing approaches: a review, Microelectronics Reliability 52 (1) (2012) 71–89.

[15] X.C. Tong, Advanced Materials for Thermal Management of Electronic Packaging, vol 30, Springer Science & Business Media, 2011.

[16] P. Singh, C.M. Tan, Uncover the degradation science of silicone under the combined temperature and humidity conditions, IEEE Access 6 (2017) 1302–1311.

[17] Y. Zhou, et al., One-component, low-temperature, and fast cure epoxy encapsulant with high refractive index for LED applications, IEEE Transactions on Advanced Packaging 31 (3) (2008) 484–489.

[18] C.-H. Lin, et al., Development of UV stable LED encapsulants, in: 2009 4th International Microsystems, Packaging, Assembly and Circuits Technology Conference, IEEE, 2009.

[19] S.C. Yang, et al., Thermally resistant UV-curable epoxy–siloxane hybrid materials for light emitting diode (LED) encapsulation, Journal of Materials Chemistry 22 (18) (2012) 8874–8880.

[20] J. Kim, B. Ma, K. Lee, Comparison of effect of epoxy and silicone adhesive on the lifetime of plastic LED package, Electronic Materials Letters 9 (4) (2013) 429–432.

[21] Y.-H. Lin, et al., Development of high-performance optical silicone for the packaging of high-power LEDs, IEEE Transactions on Components and Packaging Technologies 33 (4) (2010) 761–766.

[22] A.W. Norris, M. Bahadur, M. Yoshitake, Novel silicone materials for LED packaging, in: Fifth International Conference on Solid State Lighting, vol 5941, International Society for Optics and Photonics, 2005.
[23] M.T. Cher, P. Singh, A review on degradation physics of high-power LEDs in outdoor applications, in: 2016 IEEE 23rd International Symposium on the Physical and Failure Analysis of Integrated Circuits (IPFA), IEEE, 2016.
[24] G. Lin, Y. Zhou, N. Tran, W. Chiang, Y. He, F.G. Shi, Materials challenges and solutions for the packaging of high-power LEDs, in: Proc. 11th Int. Symp. Adv. Packag. Mater., Process., Properties Interface, Mar. 2006, p. 63.
[25] K.-J. Byeon, J.-Y. Cho, J.O. Song, S.Y. Lee, H. Lee, High-brightness vertical GaN-based light-emitting diodes with hexagonally close-packed micrometer array structures, IEEE Photonics Journal 5 (6) (December 2013). Art. no. 8200708.
[26] C.M. Tan, et al., Rapid light output degradation of GaN-based packaged LED in the early stage of humidity test, IEEE Transactions on Device and Materials Reliability 12 (1) (2011) 44–48.
[27] W. Kim, et al., Thermally conductive EMC (epoxy molding compound) for microelectronic encapsulation, Polymer Engineering & Science 39 (4) (1999) 756–766.
[28] J.-C. Huang, et al., Comparison of epoxy resins for applications in light-emitting diodes, Advances in Polymer Technology: Journal of the Polymer Processing Institute 23 (4) (2004) 298–306.
[29] H.W. Sung, H.L. Hsu, C.C. Shih, D.S. Lin, Cross-linking characteristics of biological tissues fixed with monofunctional or multifunctional epoxy compounds, Biomaterials 17 (14) (1996) 1405–1410.
[30] file:///C:/Users/HP/Downloads/Chemical%20compatibility%20of%20LEDs%20(1).pdf.
[31] J.-S. Kim, S.C. Yang, B.-S. Bae, Thermally stable transparent sol– gel based siloxane hybrid material with high refractive index for light emitting diode (LED) encapsulation, Chemistry of Materials 22 (11) (2010) 3549–3555.
[32] H. Ardebili, J. Zhang, G. Michael, Pecht, 2 - plastic encapsulant materials, in: Haleh Ardebili, Jiawei Zhang, Michael G. Pecht, in Materials and Processes for Electronic Applications, Encapsulation Technologies for Electronic Applications, second ed., William Andrew Publishing, 2019, ISBN 9780128119785, pp. 47–121, https://doi.org/10.1016/B978-0-12-811978-5.00002-X.
[33] https://www.led-professional.com/business/reports/packaged-led-market-report-from-yole-developpement-epic-forecasts-tremendous-growth-between-2012-and-2018.
[34] C.M. Tan, P. Singh, Time evolution degradation physics in high power white LEDs under high temperature-humidity conditions, IEEE Transactions on Device and Materials Reliability 14 (2) (2014) 742–750.
[35] Q. Zhang, et al., DynamicMechanical properties of the transparent silicone resin for high power LED packaging, in: International Conference on Electronic Packaging Technology & High Density Packaging, 2008, ICEPT-HDP 2008 July 28–31, 2008, pp. 1–4, https://doi.org/10.1109/ICEPT.2008.4607073.
[36] https://www.mouser.jp/pdfdocs/LumiledsWP15.pdf.
[37] K.R. Shailesh, et al., Application of accelerated life testing principles to project long term lumen maintenance of LED luminaires, in: 2012 International Conference on Emerging Trends in Electrical Engineering and Energy Management (ICETEEEM), IEEE, 2012.
[38] J. Fan, Y. Kam-Chuen, M. Pecht, Lifetime estimation of high-power white LED using degradation-data-driven method, IEEE Transactions on Device and Materials Reliability 12 (2) (2012) 470–477.
[39] https://www.osti.gov/servlets/purl/1360770.
[40] P. Singh, C.M. Tan, Degradation physics of high power LEDs in outdoor environment and the role of phosphor in the degradation process, Scientific Reports 6 (2016) 24052.

CHAPTER 2

# General failure analysis techniques for packaged LED

## 2.1 Introduction

LEDs have been used into various application fields due to their many beneficial characteristics as discussed in the previous chapter. However, the various operation conditions and requirements in different application fields can reduce the reliability of LEDs significantly owing to the possible harsh operational conditions such as high humidity, high temperature, high UV exposure, etc. [1—6]. To assess the suitability of their applications, a good knowledge of various possible failure mechanisms in different operating conditions is required, and these failure mechanisms can only be understood from detailed failure analysis methods. Usually, integration of various failure analysis approaches is needed, which include the following:

### 2.1.1 Failure conditions

The conditions at which an LED failed must be known. These include the environmental conditions, the duration at which the LED is exposed to this condition, and the progression of the degradation in this condition [7,8].

### 2.1.2 Measurements

Electrical and optical measurements of the LED are needed. These include optical output power versus drive current, optical output spectrum, and the current—voltage characteristics. Optical luminous data provide LED degradation trend and its failure time [9,10], whereas the electrical measurement provides information on its electrical connectivity and thermal degradation of its package [11,12]. It also includes significant junction parameters such as ideality factor [13—15] and reverse saturation current [16—18], which can provide useful details about the LED chip, wire bonding pads, and wires used for electrical connections in LEDs.

*Reliability and Failure Analysis of High-Power LED Packaging*
ISBN 978-0-12-822408-3
https://doi.org/10.1016/B978-0-12-822408-3.00008-3

### 2.1.3 Optical microscopy

It provides information regarding the deterioration of the LED package [19–21].

### 2.1.4 Nondestructive examination (NDE)

Nondestructive analysis plays a crucial role in LED packaging. Generally, these NDE techniques include X-ray inspection [8,22–24], scanning acoustic microscopy (SAM) [24–26], Fourier transform infrared microscopy (FTIR) [27–29], and infrared thermography [30–32]. These methods can provide information regarding the internal delamination, voids or cracks propagation, moisture ingression, etc. NDE methods can help to narrow down the failure examination areas and thereby allow detail failure analysis to be performed.

### 2.1.5 Destructive examination

After locating the defective areas in packaged LED using NDE techniques, destructive techniques are employed where LED package is disassembled/decapsulated to observe in the detail of the defective areas. In the case of LED, encapsulant is removed to observe the conditions of LED chip, wire bonding, molding surface, etc. Scanning electron microscopy (SEM) [33,34], energy-dispersive X-ray spectroscopy (EDX) [33,35,36], FTIR [27–29], and other inspection methods are widely used after destructive decapsulation. These methods help to understand the topographic, elemental as well as chemical bonding alteration occurred at the defect sites so as to propose failure causes along with suggestions to overcome these failure causes.

This chapter aims to provide an overview of the abovementioned failure analysis techniques along with their basic operation principles, advantages, and limitations. Various case studies are also presented, illuminating the usefulness of the individual analysis techniques in LEDs failure analysis. Separate sections will provide information regarding microscopic, spectroscopy, radiography, thermal, surface analysis, and ultrasound-based failure analysis investigations. Later part of this chapter deals with the decapsulation techniques and sample preparation methods used to prepare or uncover the defective areas from a packaged LED.

## 2.2 Functional and parameters examination

The reverse saturation current *(Is)*, series resistance (Rs), and ideality factor (n) of the LED junction in each LED can be computed using an extraction method developed by Tan et al. [37] to examine the impact of stress factors, such as temperature and moisture, on the electrical characteristics of LED chips during their operations, including reliability tests.

These criteria provide a wealth of knowledge about the packaged high-power LEDs that have been operated under various environmental conditions. The ideality factor of a junction is a metric for how much carrier is injected into it [38–40]. The diffusive movement of minority electrons from the p-side to the n-side and minority holes from the n-side to the p-side induces the reverse saturation current *Is* in a PN junction [16,41,42]. The reverse saturation current is determined by the carrier lifetime, electron and hole diffusion coefficients, and the energy barrier at the junction for the carriers. Since the *Is* values for LED chips are typically small, a significant shift can be observed for a small change in LED electrical characteristics, making it the most sensitive indicator for LED chip degradation. Because of its small value, it cannot be measured and must be calculated using an algorithm defined in Refs. [16,43]. As seen in Refs. [1,16,43,44], this algorithm has proven to be efficient in a variety of applications.

When a p-n junction is shunted, due possibly to moisture or ionic contamination, an electrical path is created, and current may flow through it under reverse bias, rising *Is* [16,43]. Thus, a formation of shunt path through a pn junction is usually represented by an increase in the value of *Is* alone, i.e., the values of n and Rs do not change significantly. Tan et al. [16,43] often used these criteria to investigate the health of LED chip, wire bond, or bonding condition.

Ma et al. used an in situ accelerated method and various analytic technologies to investigate the degradation behavior of 260 nm flip-chip UVC-LEDs. LEDs were stressed at a constant DC current of 20 mA, and the optical strength of the LEDs had decreased to 63% of its initial value after 150 h of service. Fig. 2.1 shows their I-V characteristics before and after the test. Their ideality factor was found to increase approximately linear with stress time, and this behavior was attributed to the formation of point defects inside and around active regions, which were responsible for the nonradiative recombination and trap-assisted tunneling [45].

Jayawardena et al. investigated the variation of ideality factor and series resistance of LEDs until they failed under various currents and temperature stress conditions, and the results are shown in Fig. 2.2. Their findings indicated that catastrophic short-circuit failure of LEDs was caused by electromigration-induced metal migration from the contact metallization layer to the semiconductor. Their hypothesis was backed up by variations in ideality factor and reverse leakage current. The significance of these critical factors in recognizing the hidden failure was highlighted in their report.

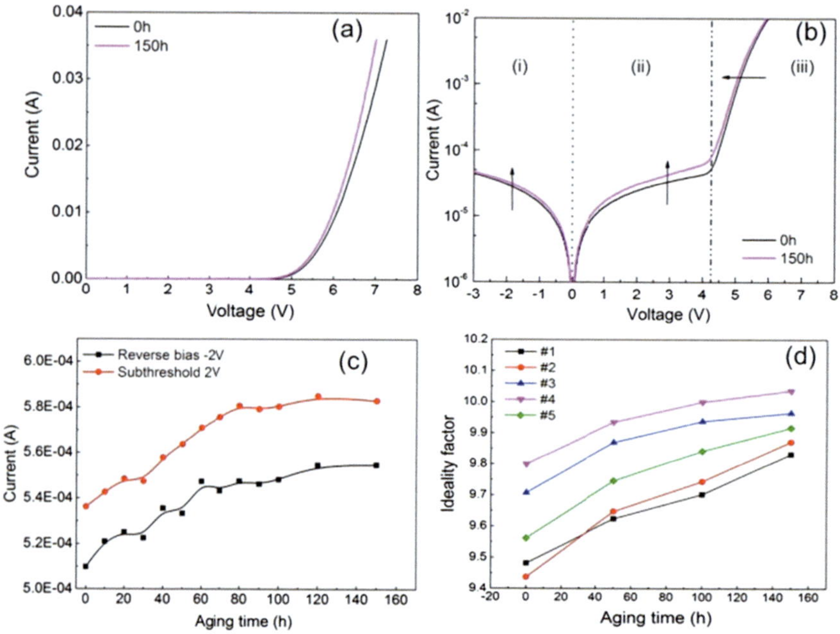


**Figure 2.1** (a) and (b) I–V characteristics of the 260 nm ultraviolet LED, measured before and after stress. In (b), region (i) is in the reverse bias; region (ii) is in the low-forward bias; region (iii) is in the bias region above the turn-on voltage; (c) The reverse leakage current and the subthreshold current at 2 V of the LED sample during stress; (d) UVC-LED 260 nm ideality factor variance during stress for different samples [45]. *(©Solid- State Electronics, Elsevier. Reprinted with permission.)*

In their study, Asiri et al. [47] looked at how electrical parameters of InGaN-based LED packages changed with age. Electrical parameters, especially package series resistance, were found to be good indicators of LED package health (i.e., remaining useful life) in their research and could potentially be sensed in real time in applications by Asiri et al.

In short, electrical measurements, especially ideality factor, series resistance, and reverse saturation current measurements, are powerful nondestructive techniques that provide a good indicator for a specific failure mechanism, allowing subsequent failure analysis steps to be more directed and thus saving analysis time.

## 2.3 Microscopic inspection

The most commonly used nondestructive failure analysis technique is optical microscopy. It can quickly identify any external defect without

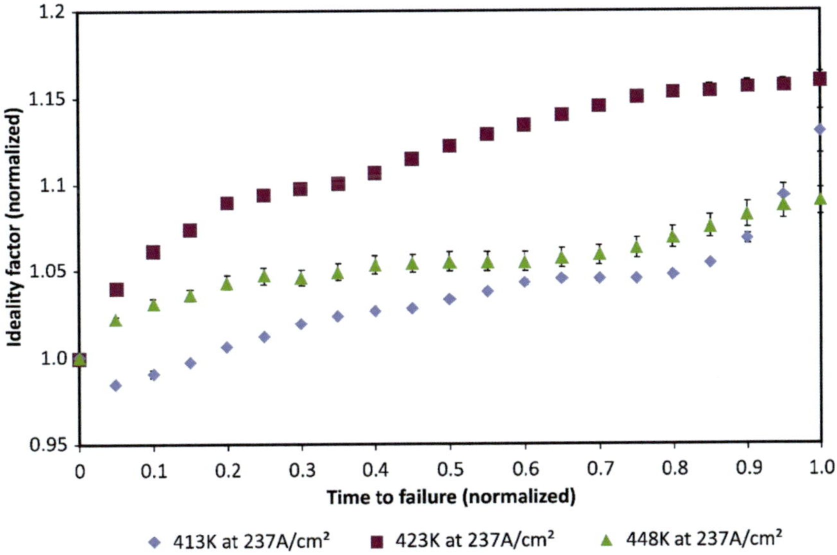


(a)

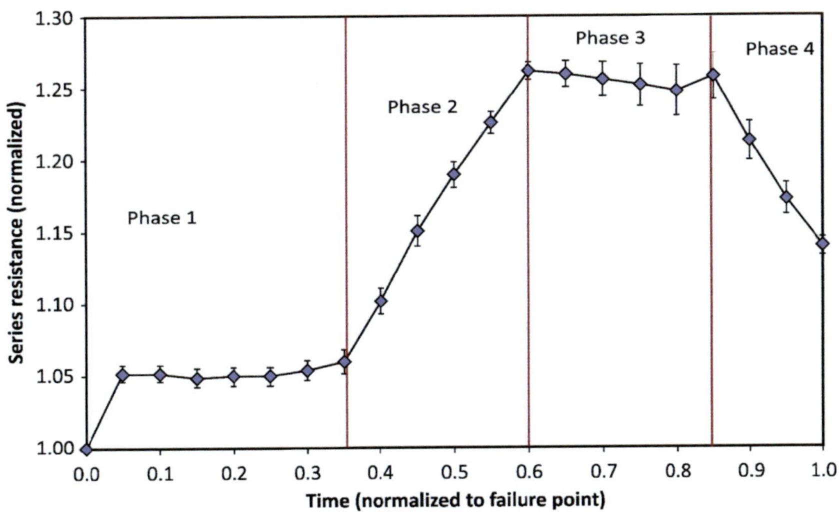


(b)

**Figure 2.2** (a) Ideality factor variation with aging for LED packages under different stress conditions and (b) series resistance variation with aging for an LED package at 423 K and 296 A/cm$^2$ [46]. *(©Microelectronics Reliability, Elsevier. Reprinted with permission.)*

requiring extensive sample preparation, and it is very simple to use. An optical microscope produces a magnified image of the object under examination using an objective lens and magnifies the image even further using an eyepiece, enabling the user to display the image with their naked eyes [19,48–50]. With the advancement in electronics, today optical microscopes are connected to image processing unit and very high-resolution images can be obtained and saved for future investigations.

Optical microscope has been widely employed for LEDs failure analysis, for the examination of encapsulant-molding part delamination, bonding pad degradation, bond wire lift-off failure, molding part surface discoloration along with crack propagation, LED chip detachment, electromigration-induced degradation at the electrode surface, etc. [19–21]. We will briefly explain some of the failure analysis case studies of LEDs using optical microscope in this section.

Leterrier et al. [51] examined the mechanical integrity of tin-doped indium oxide (ITO) thin films sputtered onto a high-temperature aromatic polyester, which is critical for the flexible display applications such as organic LEDs (OLEDs). They investigated the mechanical integrity by means of tensile experiments and employed an in situ optical microscopy to observe the defects or cracks patterns on the ITO surface as shown in Fig. 2.3. Process-induced internal stresses were found to be compressive, and cracks in ITO layers began at pin-hole defect sites. The transition from stable crack growth to unstable propagation was also visible in the optical microscope images.

Blackening-induced lumen decay is a common issue in LEDs, especially during WHTOL reliability tests. Zhang et al. [52] observed this mechanism and discussed the reasons for its failure. They discovered that blackening failure was caused by delamination between lead frame and reflector polymer composites, followed by chemical penetration, composite corrosion, and silver migration. After mechanical decapsulation, optical microscope images of LED samples aged for 500 h under WHTOL test clearly showed the morphology of the reflector composite and lead frame surfaces (Fig. 2.4).

Park et al. [53] used optical microscopy to examine the interface between the LED and the sapphire substrate. Selective laser lift-off (SLLO) technology was commonly used for the fabrication of LED display. In this work, they investigated the fundamental relationship between laser irradiation and adhesion, as these two variables have an effect on the adhesion at

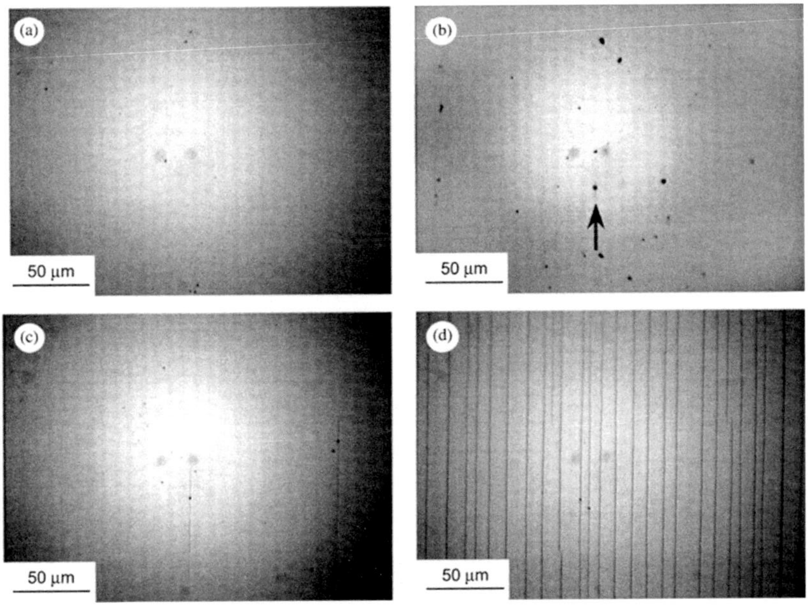


**Figure 2.3** During tensile loading, a 100 nm thick ITO series II coating cracks gradually (along the horizontal direction). Unloaded ITO (a); 1.28% strain (b); 1.42% strain (c); and 3.42% strain (d) [51]. *(©Thin Solid Films, Elsevier. Reprinted with permission.)*

the LED and sapphire substrate interface. Fig. 2.5 shows bottom-view microscopic images of the interface between the LED and sapphire substrate and the top-view microscopic images of the sapphire substrate, before and after SLLO with different laser power densities.

Singh and Tan observed various package-related failure mechanisms in high-power LEDs using an optical microscope. When blue and white high-power LEDs were stressed at high temperature and under high humidity conditions under ON conditions, delamination at the molding and encapsulant interface was observed [54]. Wide cracks were formed at the interface during the test, as shown in optical micrographs in Fig. 2.6. The difference in the coefficient of thermal expansion (CTE) of the molding and encapsulant was found to be the cause of such delamination.

In a separate study [55], Singh et al. used an optical microscope to observe the discoloration of LED molding element under harsh conditions, as shown in Fig. 2.7. They employed 3D optical microscope to obtain 3D image of the damaged surface and evaluated the depth of the cracks caused by high temperature and moisture from the ambient.

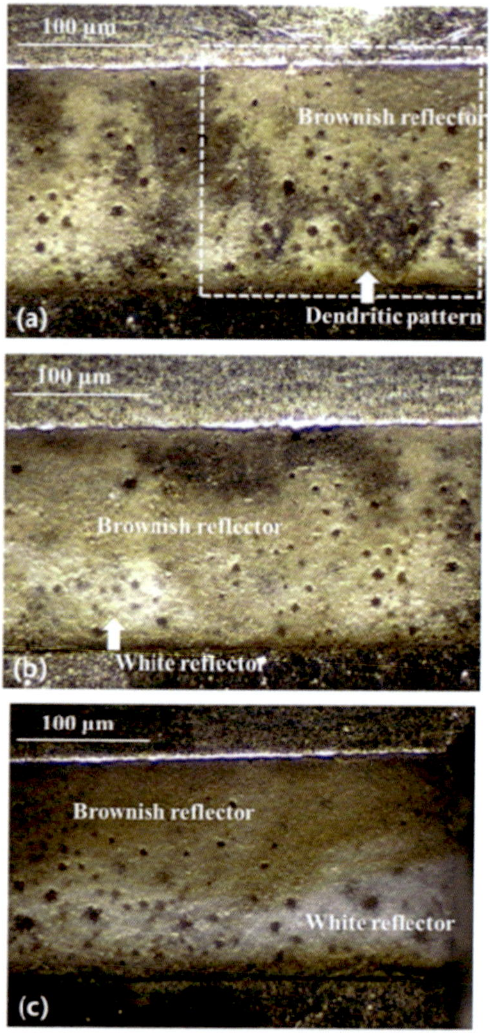


**Figure 2.4** After mechanical decapsulation, morphology of reflector composite and lead frame surfaces. Under the WHTOL exam, the sample had been aged for 500 h [52]. *(©Microelectronics Reliability, Elsevier. Reprinted with permission.)*

Understanding and checking the failure mechanisms of LEDs using an optical microscope have proved to be very effective and accurate. Optical microscopy may show the failure mechanisms in LEDs, yet, spectroscopic analysis is needed to further examine and gain a deeper understanding of the extent of material properties changes after the failure of molding, encapsulant, solder pads, or LED chip.

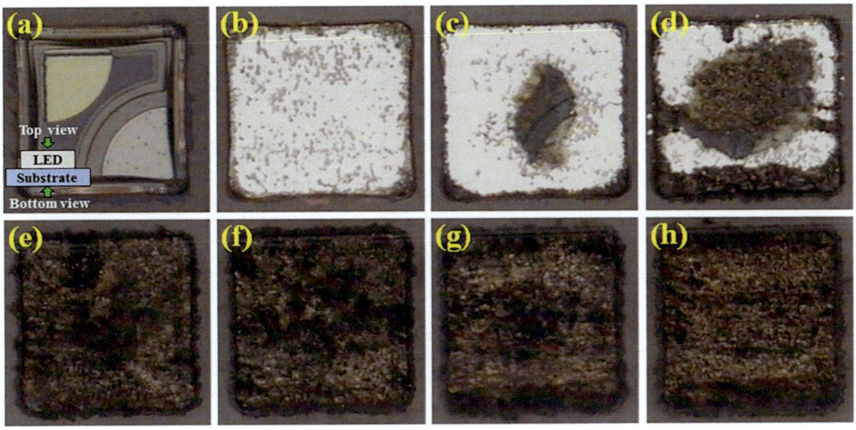


(a)

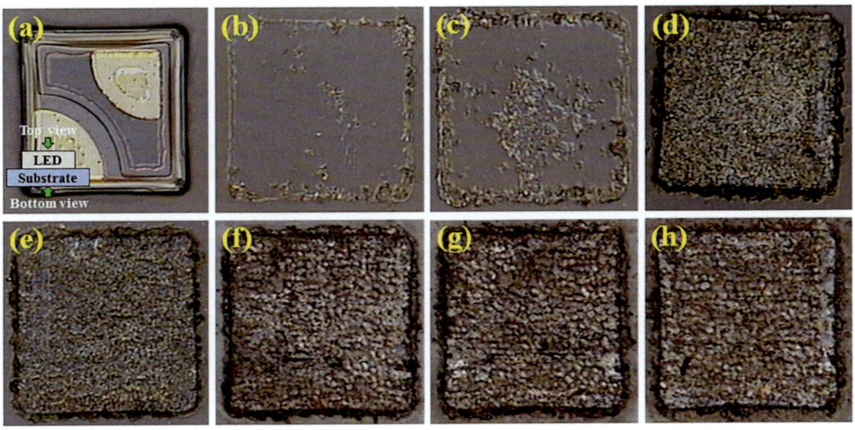


(b)

**Figure 2.5** (i) Microscopic photographs of the LED–sapphire substrate interface from the bottom. (a) Before and after SLLO with laser power densities of (b) 64, (c) 70, (d) 76, (e) 83, (f) 89, (g) 95, and (h) 102 kW/cm$^2$. (j) Micrographs of the sapphire layer in top view. (a) Before and after separation of the LED by SLLO with laser power densities of (b) 64, (c) 70, (d) 76, (e) 83, (f) 89, (g) 95, and (h) 102 kW/cm$^2$ [53]. *(*©Applied Surface Science, *Elsevier. Reprinted with permission.)*

## 2.4 Spectrometric examination

Optical microscope can only provide the topographical and appearance change at the material surface before and after the failure, and spectroscopic examinations are required to understand the change in the materials at

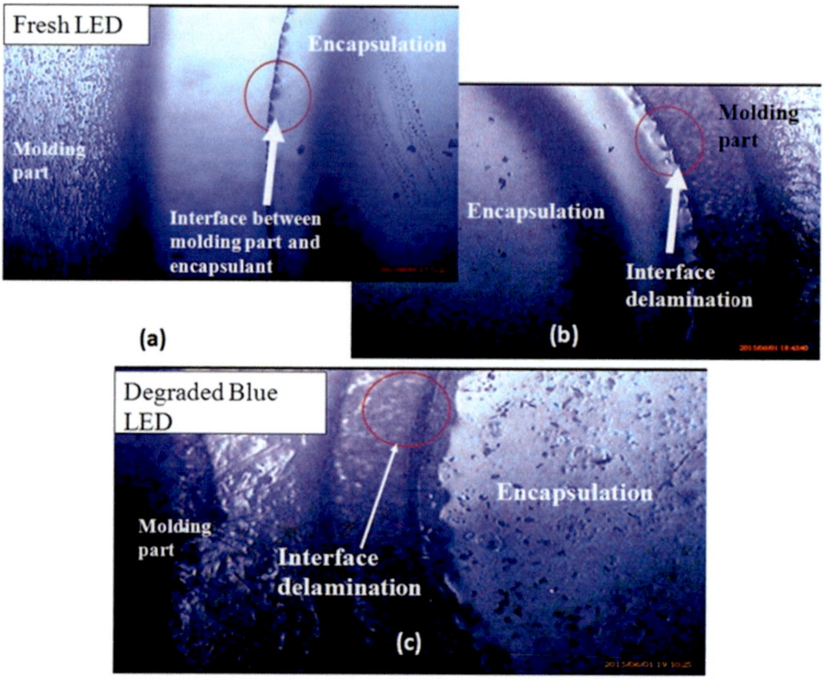


**Figure 2.6** Optical microscope images at 5 x magnification showing delamination caused by thermal expansion variations in (b) white LEDs and (c) blue LEDs as compared to (a) fresh LED [54]. *(©IEEE Access, IEEE. Reprinted with permission.)*

molecular level. Fourier transform infrared (FTIR) [56,57] and Raman spectroscopy [58,59] are the two most common techniques used for material analysis. They are readily available and provide rapid results for the samples under investigation, without much sample preparation.

Generally, LEDs' material changes are related to thermal decomposition, hydrolysis, thermal aging, etc. [55,60]. Raman spectroscopy is not suitable for hydrolysis studies (since water is a weak laser scatterer) [61] along with the fluorescence interference as fluorescence of impurity or material itself can hide Raman spectrum, which is not an issue with FTIR spectroscopy. Therefore, FTIR is more commonly used in LED failure analysis.

FTIR spectroscopy is a measurement of the interaction of infrared radiation with matter, which depends on the reflectance, absorbance, or transmittance of the infrared light in a material at distinct frequencies, which directly correspond to the vibrational frequency of the molecular bonds of a material [56,57].

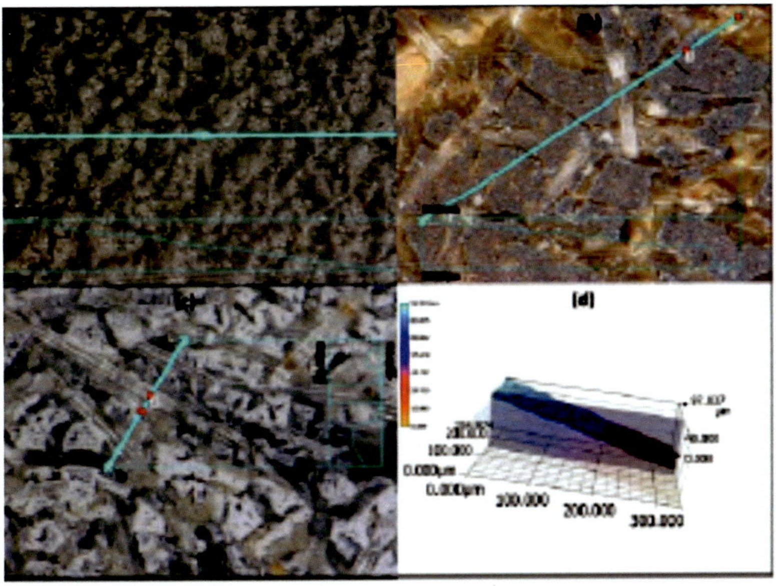


**Figure 2.7** At 1000 x magnification, 2-D microscopic images of the surface of the silicone molding component for (a) fresh LED, (b) white degraded LED, and (c) blue degraded LED. (d) Images of the depth profile of a fresh LED surface as a depth measurement example. The color spectrum represents the height of cracks in various locations [55]. *(©IEEE Access, IEEE. Reprinted with permission.)*

Polycarbonates (PC) are widely used in LEDs for packaging, and it can undergo various material transformations under harsh environments such as high temperature, high moisture, UV radiation, etc. [62,63]. Rivaton et al. investigated the impact of temperature on PC degradation and discovered that the oxidation reaction that causes PC degradation can be divided into two types, namely chain and ring oxidation [64]. They carried out accelerated thermal aging at 170°C, and their FTIR results revealed that dicarboxylic acids formed in the ring were thermally oxidized and formed anhydrides, as evidenced by the appearance of infrared absorption bands at 1860 and 1840/cm. The appearance of infrared absorption bands at 1724 and 1690/cm was due to aliphatic and aromatic ketones as evidence of the side-chain formation as shown in Fig. 2.8 [65]. According to Meher et al., the oxidation reaction in bisphenol A polycarbonate (BPA-PC) can be easily verified by analyzing the specimens' FTIR-ATR spectra. Any rise in absorption rate at 1690, 1840/1860, and 1713/cm is linked to cyclic anhydrides, aromatic ketones, and cyclic anhydrides formation, and these may be an evidence of the thermal oxidation reaction [64,66].

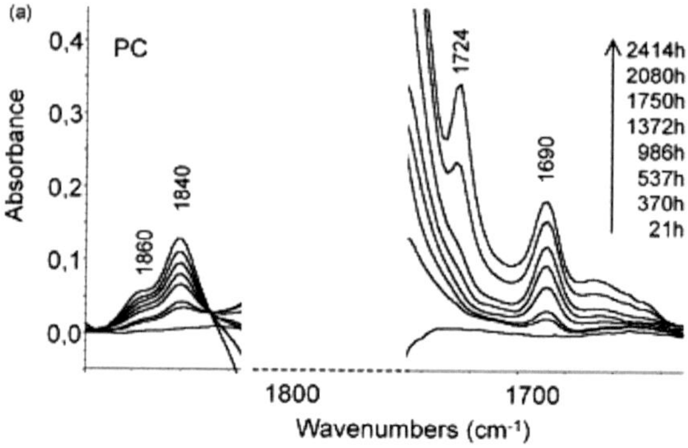


**Figure 2.8** Changes in the carbonyl region of thermo-oxidized PC film IR spectra at 170°C [65]. Infrared absorption bands at 1860 and 1840 cm1 and presence of infrared absorption bands at 1724 and 1690/cm due to aliphatic and aromatic ketones characterizing anhydrides. *(©Optical Materials, Elsevier. Reprinted with permission.)*

Another important LED packaging material is PMMA due to its low cost and application as a secondary lens. Yellowing of PMMA material under high-temperature conditions is observed by Lu et al. [67], and they attributed this yellowing to thermal aging of PMMA when subjected to thermal exposure of 150°C. This finding was verified using FTIR as shown in Fig. 2.9. Considerable rise in oxidation peaks was observed along with −COOH peak, verified the oxidation and thermal aging of PMMA after the high-temperature stress test.

Due to various advantages such as low impurity levels, good moisture resistance, high optical clarity, and low modulus, poly dimethyl siloxane, also known as silicone, has replaced PC and PMMA as a more superior packaging material for LEDs in recent years [68,69]. Singh et al. examined the efficiency of PDMS in high-power blue and white LEDs under extreme temperatures and humidity. They observed that silicone can also undergo various degradation mechanisms such as thermal oxidation, hydrolysis, condensation, and aging over their lifetime under high moisture and temperature conditions when used as a package material in high-power LEDs [55,70]. Thermal aging of silicone will be accompanied by an increase in Si-$CH_3$ peak, hydrolysis process can be observed by the rise in−OH peak, −C=O can be a by-product originating from a thermal oxidation reaction, and decrease in Si-O or Si-$CH_3$ peaks can provide vital information regarding the hydrolysis as well as cross-linking phenomenon of silicones. FTIR is used to verify the abovementioned mechanisms as shown

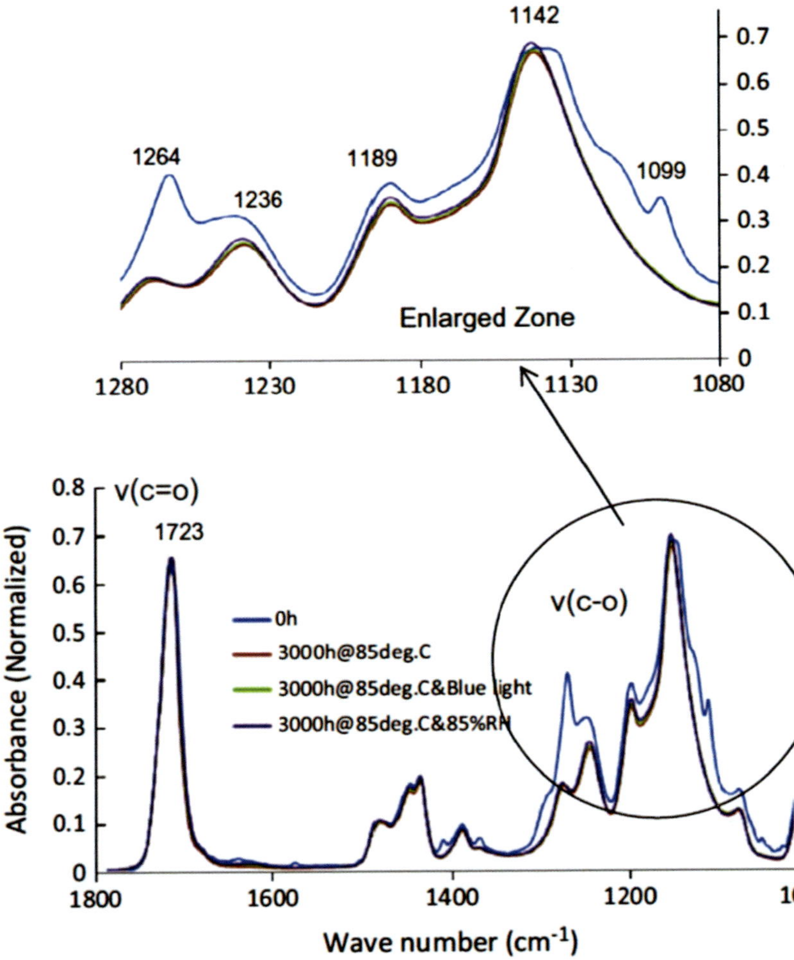


**Figure 2.9** IR spectra of aged PMMA under various conditions [67]. *(*©Optics letters, Elsevier. *Reprinted with permission)*

in Fig. 2.10. In-depth silicone degradation study and its impact on LEDs light intensity will be discussed in later chapters.

As spectroscopy techniques can provide the chemical change occurring in the LED packaging material under environmental stress, it is an important failure analysis tool. However, decapsulation of LEDs is generally required to perform such spectroscopic analysis, which is considered as destructive, unless the analysis is done on the outside of the package.

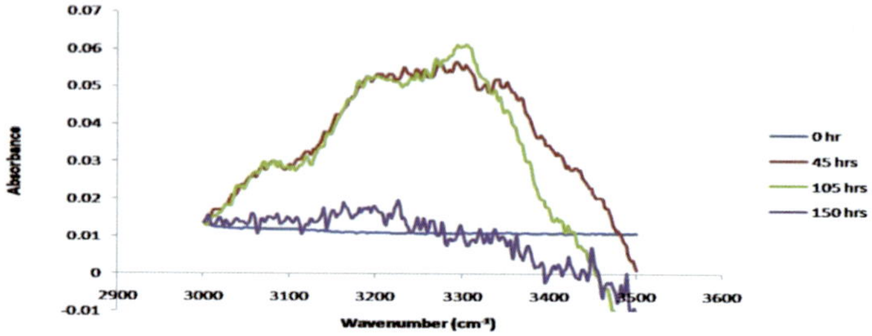


(a)

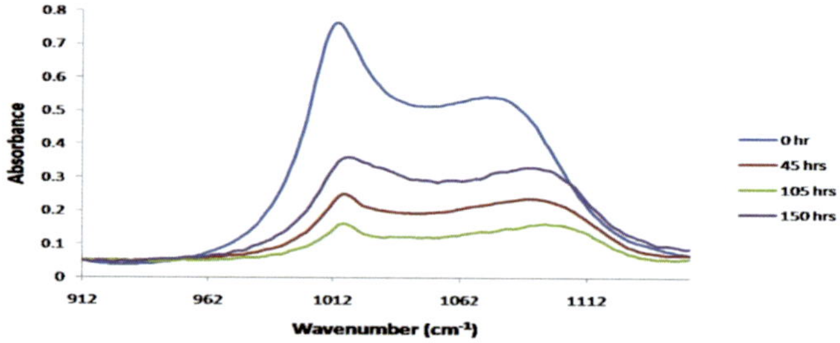


(b)

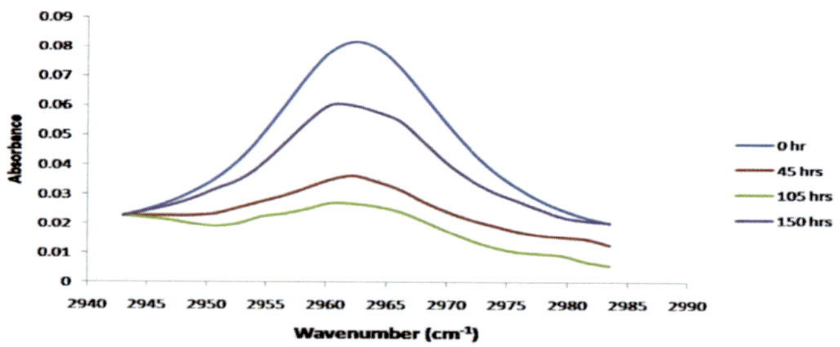


(c)

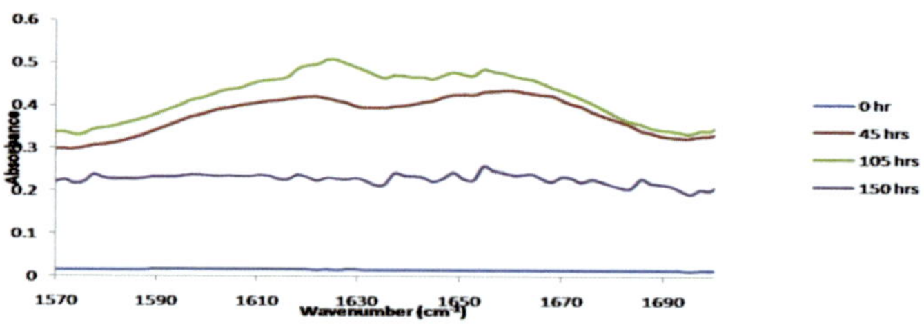


(d)

**Figure 2.10** Comparison of FTIR peaks for new and deteriorated samples. (a) —OH peak variation, (b) Si-O-Si peak variation, (c) Si-$CH_3$ peak variation, and (d) —C═O peak variation [70]. (©Optical Materials, *Elsevier. Reprinted with permission.)*

Another important nondestructive technique that can help to locate the defects in the internal parts of the LEDs is radiography examination.

## 2.5 Radiography examination (X-ray CT)

X-ray microscopy provides a nondestructive assessment of internal defects, damage, and degradation in most of the microelectronic devices including packaged LEDs. X-ray illumination on a sample provides images based on the density of the various materials used for LED packaging and these images allow characterization of die attachment voiding, solder pooling, die shifting, or wire bond breakage inside the LED package [71].

In another work, Tan et al. investigated the white packaged LEDs degradation caused by accelerated humidity testing [72]. They employed X-ray microscopy as shown in Fig. 2.11, which shows the left and right cross-sectioning of the failed LED sample; however, no abnormality was observed from these images. Tan et al. used X-ray microscopy to confirm the cause behind the change in electrical parameters such as reverse saturation current, ideality factor, and series resistance.

Similarly, others have employed X-ray technique for in-depth analysis of LED failure and observed wire bond breakage at the wedge bond, which caused sudden LED failures [73].

Due to the beam hardening effect, X-ray CT scanning will display a 3D image of the LEDs (beam hardening generally occurs when an object consists of different parts having different attenuation coefficients to X-ray and the objects under scanning tends to appear denser at the edges than at the center, creating artifacts); however, it was unable to detect delamination

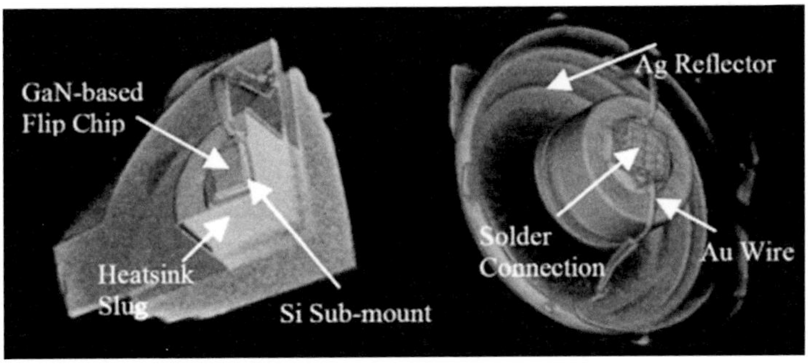


**Figure 2.11** Three-dimensional X-rays of an LED sample displaying cross-sectioning (left) and complete (right) views [72]. *(©IEEE TDMR, IEEE. Reprinted with permission.)*

between silicone and lead frame or defects in low-density materials, necessitating the use of acoustic microscopy.

## 2.6 Thermography examination

Thermography examination is another nondestructive technique generally used for void or defects identification on a large driver board for driving LEDs. Any void or defect will result into a thermal hot spot and the infrared energy emitted from it can be detected. Thermography converts the infrared energy into temperature and displays the image of the temperature distribution [74,75].

Infrared thermography is generally used to compare the temperature rise of an LED after the failure. As LEDs use heat sink to dissipate heat, which is attached to LED chip via solder paste, the solder paste can lose its thermal conductance, or voids can accumulate, which can lead to inefficient heat dissipation in the LEDs, and a rapid temperature rise in the packaged LED will result. All these can be detected using infrared thermography technique. Infrared thermography can also reveal the temperature change pattern due to irregularity or degradation or varying amount of phosphor, which is used as a down-conversion material in LEDs to generate white light.

Tan et al. looked into the physical limitations of phosphor layer thickness and concentration for white LEDs [76], and they discovered that these variables are constrained by light ray scattering, with blue light rays having a higher percent decrease than yellow light rays. As shown in Fig. 2.12, they used IR thermography to compare the temperature rise patterns of different samples with different phosphor particles. Since the conversion efficiency of blue to yellow light is not perfect, the heat produced during the conversion increases as more phosphor particles are added, resulting in a steady rise in the temperature of the phosphor layer, as shown in Fig. 2.12.

Singh et al. employed infrared thermography to investigate the LED failure mechanism when high-power white LED is subject to high temperature, moisture, and electrical stress [**89**]. They found that the overall temperature of LED decreases initially for first few hours and later increases back as shown in Fig. 2.13. They correlated this temperature trend with the material degradation and understand the corresponding LED failure mechanism.

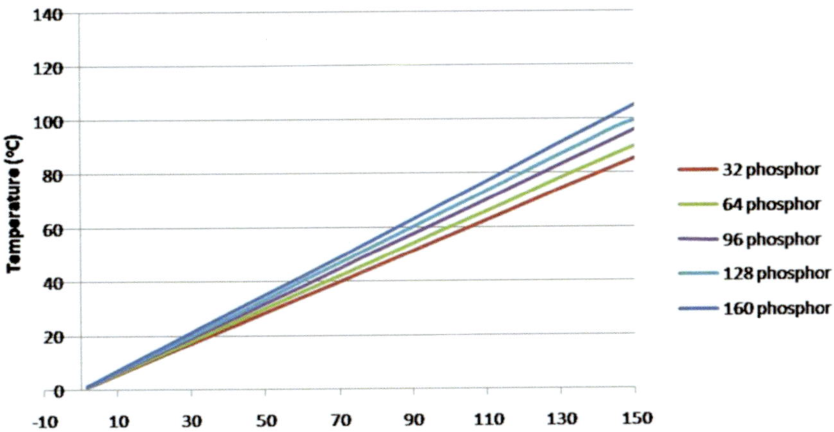


**Figure 2.12** Comparison of temperature increase between different samples with different phosphor particle numbers [76]. Legend shows the different samples with increasing number of phosphor particles, which will convert blue light incident on them and generate yellow light. *(©Scientific reports, Nature. Reprinted with permission.)*

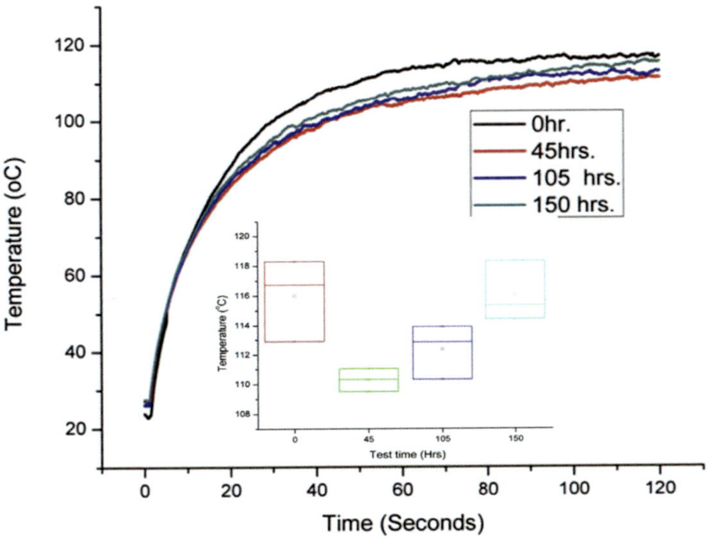


**Figure 2.13** Temperature rises of various LEDs tested over various time periods are compared. The improvement in the steady-state temperature of LEDs compared to the fresh sample is depicted in the inset [70]. *(©Optical Materials, Elsevier. Reprinted with permission.)*

As observed from these studies, infrared thermography can detect the thermal hotspots generated in the LEDs' inner structure, which will lead to overall temperature rise in the LEDs. Although this is a very good nondestructive technique for detecting failure sites, when there is stacking of different layers within the LEDs, it is difficult to evaluate the faulty area with this technique alone. For example, LED is usually placed on the heat sink with a die attachment layer, and phosphor layer is present on the LED chip along with many other parts such as bond pad and wire bonding. Thus, decapsulation of the faulty or degraded LEDs is generally needed to determine the exact failure spots at various areas inside the LEDs.

## 2.7 Scanning electron microscope analysis techniques

Scanning electron microscope (SEM) is a common tool in science and engineering for analysis. When it is used with energy-dispersive X-ray (EDX), it can help to generate more information about the sample along with topographical changes. Its resolution is well beyond optical microscope [77,78].

SEM images are generated by scanning a sample with a high-energy electron beam. Secondary electrons, backscattered electrons, and characteristic X-rays are generated as the electrons interact with the sample [79]. Different detectors capture these signals to produce images, which are then projected on a computer screen. Depending on the accelerating voltage and the density of the sample, the electron beam penetrates the sample to a depth of a few microns as it reaches the surface. This interaction within the sample produces a variety of signals, including secondary electrons and X-rays.

SEM-EDS has been widely employed in the failure analysis of LEDs. Heat dissipation is an issue for the poor reliability of LEDs and lower heat dissipation substrates or die attachment materials are the reasons for such poor reliability. Horng et al. used a nondestructive thermal resistance test method to locate the failure sites of sapphire-based InGaN LEDs [80]. The blistering in the attaching layers shown in Fig. 2.14 has thermal resistance around 4.4 K/W, and these layers were found to reduce the reliability of the overall LED package.

Liu et al. [81] used SEM to understand the interfacial reactions and failure modes of flip chip LED solder joints on Cu with electroless Ni/immersion gold (ENIG) and with organic solderability preservatives (Cu-OSP) surface finishes, respectively. The microstructures of the solder after

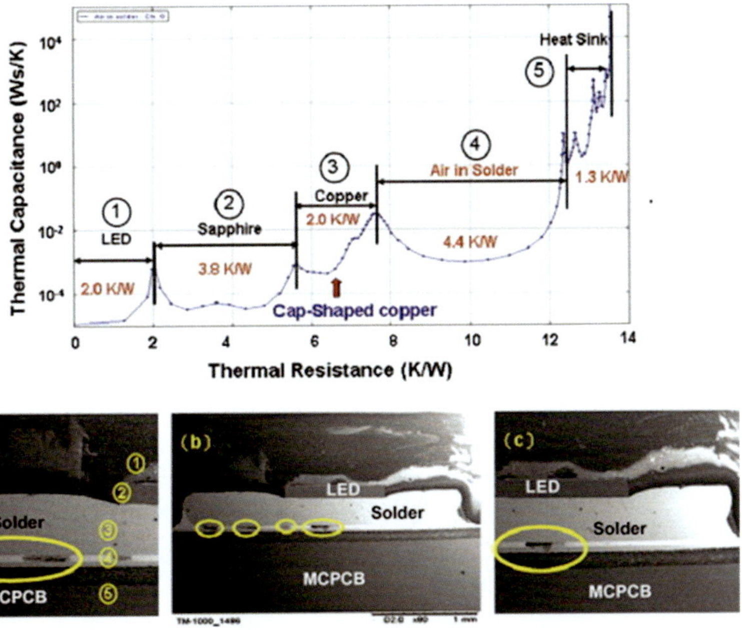


**Figure 2.14** Thermal resistance study and cross-section pictures of an LED with blisters in the solder connecting layer failure [80]. *(©Elsevier. Reprinted with permission.)*

600 h of aging at 85°C on both surfaces are shown in Fig. 2.15. The study discovered that although the interfacial reactions in the two systems vary, the failure mechanisms of the two types of solder joints during the shear test are identical. Liu et al. used various other tools along with SEM images to understand the cause behind the different failure mechanism in two systems and more details can be found in Ref. [81].

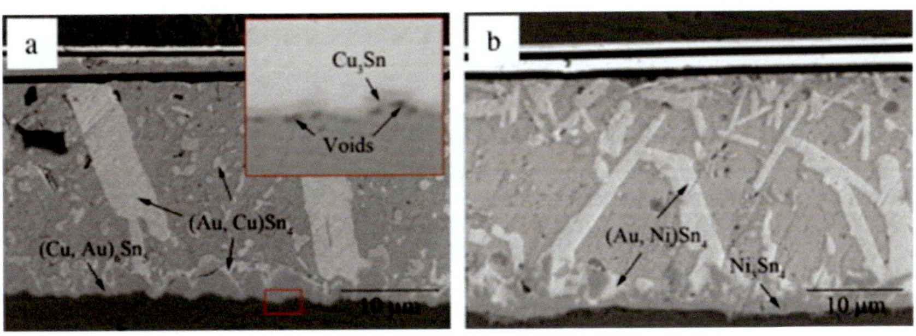


**Figure 2.15** After being aged for 600 h at 85°C, the microstructure and EDS study of the solder joints on ENIG and Cu-OSP (a) the solder joint on Cu-OSP and (b) the solder joint on ENIG [81]. *(©Elsevier. Reprinted with permission.)*

Ma et al. [45] investigated the deterioration of 260 nm ultraviolet light emitting diodes (UVC-LEDs) when stressed with a steady DC current of 20 mA in 2019. After 150 h of test, the optical power was reduced to 63% of its initial value, and the results showed that the increase in current for the reverse bias region and the low forward bias region during stress was due to an increase in defect-assisted carrier tunneling; electromigration of contact metal may have created a current leakage channel, which degraded the optical power of LEDs. SEM and EDS, as well as other failure analysis techniques, showed that the concentration of hydrogen in the p-GaN layer decreased from $4.5 \times 10^{17}$ atoms/cm$^3$ for unstressed samples to $6.0 \times 10^{16}$ atoms/cm$^3$ for stressed samples, while the Mg concentration remained unchanged, suggesting that the hydrogen was dissociated from the Mg-H complex during the initial phase as shown in Fig. 2.16.

The above case studies are just a few examples out of wide number of failure analysis studies using SEM-EDS. SEM-EDS has proven its worth as a very powerful tool to have a more in-depth knowledge of the failure mechanisms. While SEM-EDS is very useful in failure analysis, it has some drawbacks, such as poor peak energy resolution, low peak to background ratio, input signal rate limitation due to pulse processing requirements. One of the main limitations is that SEM-EDX is a destructive technique where analysts have to decapsulate the LED to investigate the inner failures and sample preparation is generally required in order to increase the surface

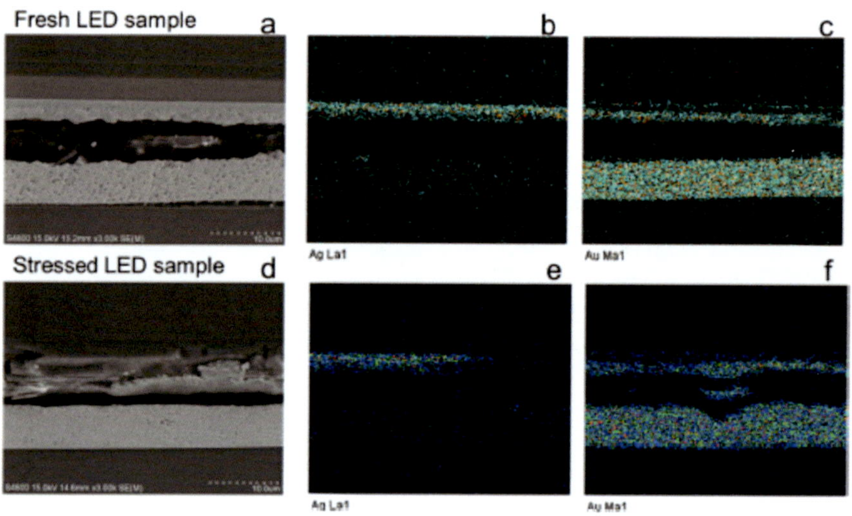


**Figure 2.16** The LED chip and die attachment region's SEM images and EDX mapping [45]. *(©Elsevier. Reprinted with permission.)*

conductivity so that electrons from the electron beam can be conducted away from the surface. The latter limitation can be overcome with the environmental SEM where low vacuum level can be accommodated.

## 2.8 Scanning acoustic microscopy examination (SAM)

Scanning Acoustic Tomography (SAT), also known as Confocal-Scanning Acoustic Microscope (C-SAM), is a technique for analyzing materials by measuring the energy and speed of a reflected ultrasonic wave from a sample. As acoustic signals are transmitted via pure water and reach the interfaces of various materials, some of the signals are mirrored and some are penetrated. SAT will detect irregularities at the interfaces using the reflected acoustic signal [82–84].

Interface delamination in LEDs after WHTOL reliability is an important failure mechanism. Zhang et al. in their work investigated the effect of WHTOL test on the LED package reliability where interface between lead frame and reflector polymer was of the prime interest [85]. They used different gates to observe the delamination before and after the test, and they discovered that a vertical delamination between the lead frame and composite was already present before the WHTOL test, and moisture may have entered the LED package along this direction, as shown in Fig. 2.17. These observations were later verified using the SEM images.

Singh et al. employed C-SAM to verify the presence of moisture inside the die attachment layer of the LED package [44]. In this work, they investigated the underlying cause of the recovery of LED lumen degradation after an early rapid lumen degradation. They used finite element modeling (FEM) to discover moisture absorption by die attachment material, which sucks the moisture from the silicone, resulted in the recovery of lumen degradation. However, once moisture reaches the inside of the die attachment, there is no other nondestructive method to verify it except C-SAM. The results are shown in Fig. 2.18.

Tiny voids (trapped bubbles) at the mold compound/die paddle interface and the LED die/die attachment interface are visible in the acoustic image of Fig. 2.18. A square in the center of the LED package easily identifies the die attachment location. The C-SAM result for the LEDs tested for 48 h (Fig. 2.18(a)) shows void (dark area) at the outer rings of the LEDs, indicating that moisture is only present in the molding compound. When the LEDs were tested till 550 h (Fig. 2.18(b)), the entire die attachment region was filled with black spots, showing that moisture

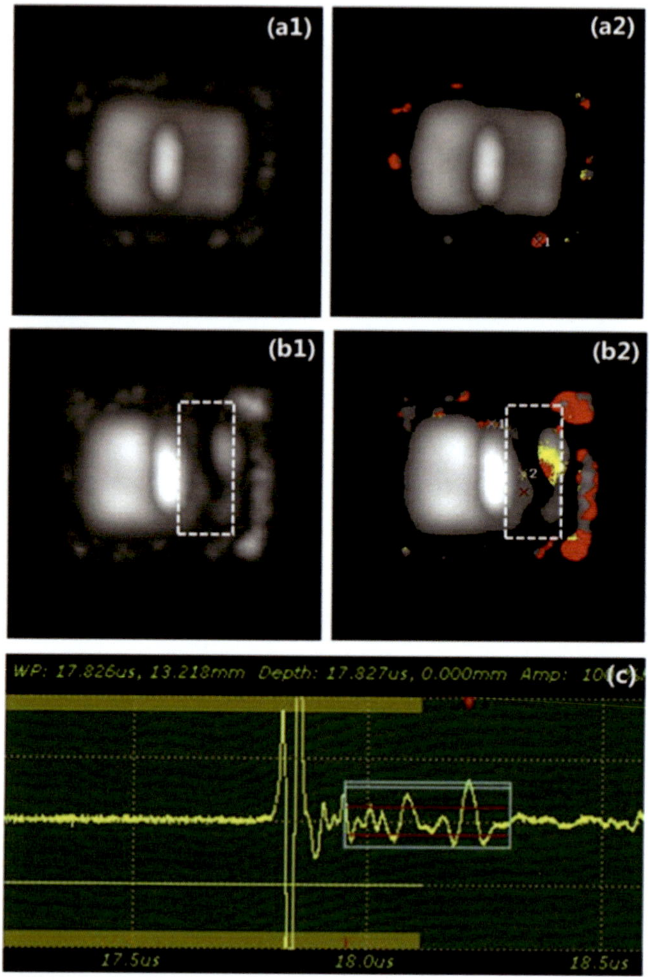


**Figure 2.17** SAM results of (a1) the data gate image before WHTOL test, (a2) the phase gate image before WHTOL test; (b1) the data gate image after WHTOL test, (b2) the phase gate image after WHTOL test, (c) A-SAM result of location 1 in WHTOL test (b2) [85]. *(©Elsevier. Reprinted with permission.)*

had penetrated the die attachment beneath the LED chip, leading to LED die/die attachment delamination. These findings are in line with the results of their simulations. This case will be further discussed in depth in the later chapters.

The above case studies demonstrate the capability of C-SAM in failure analysis of LEDs, and it is nondestructive. For an overall assessment of

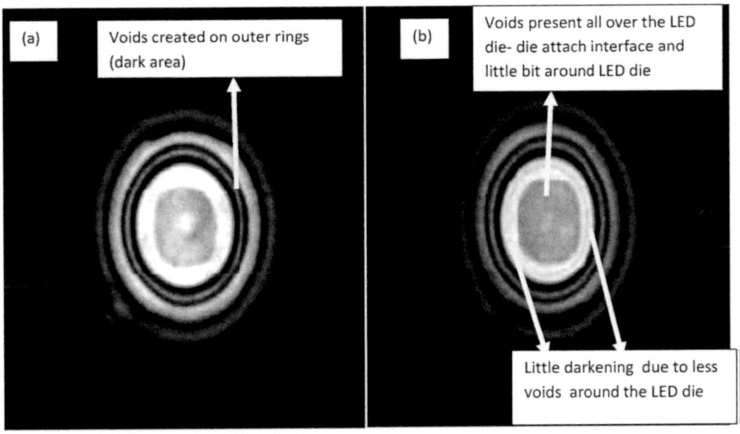


**Figure 2.18** C-SAM of a flipped LED (a) before and after initial degradation (tested for 48 h vs. tested for 550 h). After 550 h of testing, the die attachment interface with the LED die was completely blackened, but the area outside the LED die was not as darkened [44]. *(©Elsevier. Reprinted with permission.)*

failure mechanism, decapsulation is a must whereby the encapsulation or molding part is separated from the package to investigate the LED chip condition, phosphor discoloration, wire bond, or die attachment condition.

## 2.9 Internal inspection via decapsulation

Decapsulation [86–88] is the method of removing the protective layer/coating from a semiconductor die and/or any other chip used in a Plastic Encapsulated Device. This is a complex and delicate process because bond wires and internal components are extremely fragile. Mistakes made in decapsulation can create unwanted defects that lead to misleading failure analysis.

Mechanical drilling and chemical decapsulation are the two most common decapsulation methods used in the electronics industry. Mechanical drilling, however, appears to leave debris behind and isn't very accurate or delicate, so chemical etching is typically recommended.

Since LEDs have a multilayer coating and defects can occur in the inner sections of the devices, they are difficult to reach. The packaging materials must be removed in order to examine the failed parts [88]. One of the most important techniques for LED failure analysis is chemical decapsulation. Using suitable reagents to dissolve the lens and silicone encapsulants while keeping the chip, lead frame, gold wire, and other components in place, one may visually inspect or microscopically analyze the inner parts of LEDs. As shown in Ref. [72], the dissolution rate of polycarbonate (PC) lenses and silicone encapsulants using different reagents at different times is different. The mixture of $CH_2Cl_2$ and $C_6H_4(CH_3)_2$ is found to be the best choice. Fig. 2.19 shows the decapsulated samples where silicone gel with lens before and after discoloration is shown [72]. It was found that silicone gel with separate lens was harder to decap as compared to sample with lens in the silicone gel.

Chemical decapsulation can eliminate the encapsulation materials, but it can also destroy the packaging materials, and the failure sites can be lost if the failure is located within the packaging materials. Thus, a precise decapsulation method is needed with an optimum recipe, including the optimum chemical concentration and optimum etching time and correct temperature. Manoharan et al. [88] demonstrated successful decapsulation method for silicone-based encapsulant. Three-step process was used, and it starts with the removal of fraction of encapsulant compound using nitric acid. This is followed by removing most of the molding compound using a mixture of sulfuric acid and nitric acid and a postdecapsulation process for debris removal. Acetone wash is used for the postdecapsulation process.

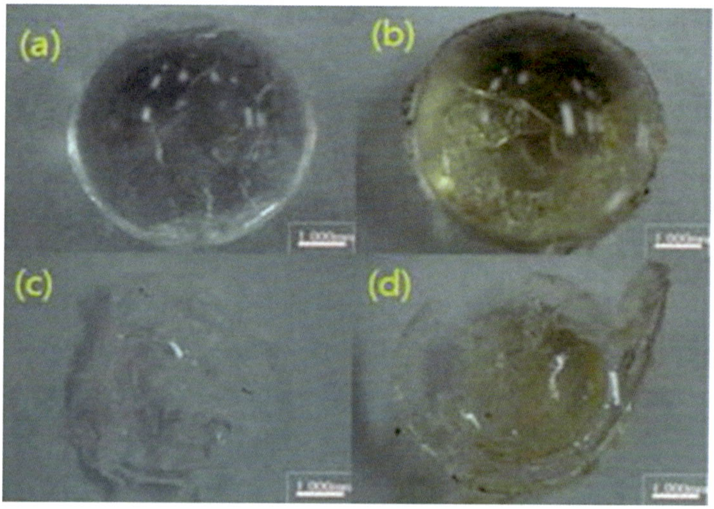


**Figure 2.19** Photographs of a decapsulation kit taken with a microscope. (a) Reference silicone gel with lens, (b) silicone gel discoloration with lens, (c) silicone gel reference from separate lens, and (d) silicone gel discoloration from separate lens [72]. *(©Elsevier. Reprinted with permission.)*

After decapping the encapsulant, failure analysis of LEDs employs techniques such as SEM, EDS, FTIR, etc., to observe the topographical and material change in the LED chip, encapulant, molding part, bond pad, etc.

## 2.10 Summary

Failure analysis is an important part of understanding the failure mechanisms of LEDs and such understanding can help to rectify the causes of failure to improve the reliability and efficiency of LEDs. This chapter discusses various failure analysis techniques for packaged LED. Both destructive and nondestructive techniques are discussed and their advantages and limitations are stated. Various microscopy techniques are mentioned along with citations from various published research for readers to understand them in detail.

## References

[1] P. Singh, C.M. Tan, Degradation physics of high power LEDs in outdoor environment and the role of phosphor in the degradation process, Scientific Reports 6 (1) (2016) 1–13.

[2] A. Zibold, M. Dammann, R. Schmidt, H. Konstanzer, M. Kunzer, Influence of air pollutants on the lifetime of LEDs and analysis of degradation effects, Microelectronics Reliability 76 (2017) 566–570.

[3] P. Lall, H. Zhang, L. Davis, Assessment of lumen degradation and remaining life of LEDs using particle filter, in: International Electronic Packaging Technical Conference and Exhibition, vol 55751, American Society of Mechanical Engineers, July 2013. V001T04A024.

[4] J. Huang, D.S. Golubović, S. Koh, D. Yang, X. Li, X. Fan, G.Q. Zhang, Rapid degradation of mid-power white-light LEDs in saturated moisture conditions, IEEE Transactions on Device and Materials Reliability 15 (4) (2015) 478–485.

[5] M. Yazdan Mehr, W. Van Driel, F. De Buyl, K. Zhang, Study on the degradation of optical silicone exposed to harsh environments, Materials 11 (8) (2018) 1305.

[6] R. Wilcoxon, J. Petroski, LEDs in harsh environments, in: Thermal Management for LED Applications, Springer, New York, NY, 2014, pp. 499–518.

[7] N. Narendran, Y. Gu, J.P. Freyssinier, H. Yu, L. Deng, Solid-state lighting: failure analysis of white LEDs, Journal of Crystal Growth 268 (3–4) (2004) 449–456.

[8] G. Lu, S. Yang, Y. Huang, Analysis on failure modes and mechanisms of LED, in: 2009 8th International Conference on Reliability, Maintainability and Safety, IEEE, July 2009, pp. 1237–1241.

[9] J. Huang, D.S. Golubović, S. Koh, D. Yang, X. Li, X. Fan, G.Q. Zhang, Lumen degradation modeling of white-light LEDs in step stress accelerated degradation test, Reliability Engineering & System Safety 154 (2016) 152–159.

[10] J. Huang, D.S. Golubović, S. Koh, D. Yang, X. Li, X. Fan, G.Q. Zhang, Degradation modeling of mid-power white-light LEDs by using Wiener process, Optics Express 23 (15) (2015) A966–A978.

[11] G. Meneghesso, S. Levada, E. Zanoni, S. Podda, G. Mura, M. Vanzi, I. Eliashevich, Failure modes and mechanisms of DC-aged GaN LEDs, Physica Status Solidi (A) 194 (2) (2002) 389–392.

[12] M. Meneghini, S. Podda, A. Morelli, R. Pintus, L. Trevisanello, G. Meneghesso, E. Zanoni, High brightness GaN LEDs degradation during dc and pulsed stress, Microelectronics Reliability 46 (9–11) (2006) 1720–1724.
[13] H. Chen, A. Keppens, P. Hanselaer, Y. Lu, Y. Gao, R. Zhuang, Z. Chen, Failure analysis of electrical-thermal-optical characteristics of LEDs based on AlGaInP and InGaN/GaN, Semiconductors 46 (10) (2012) 1310–1315.
[14] M. Meneghini, L.R. Trevisanello, S. Levada, G. Meneghesso, G. Tamiazzo, E. Zanoni, U. Straus, Failure mechanisms of gallium nitride LEDs related with passivation, in: IEEE InternationalElectron Devices Meeting, 2005. IEDM Technical Digest, IEEE, December 2005, p. 4.
[15] V. Wittpahl, Y.Y. Liu, D.S.H. Chan, W.K. Chim, J.C.H. Phang, L.J. Balk, K.P. Yan, April). A degradation monitor for the light output of LEDs based on cathodoluminescence signals and junction ideality factor, in: Proceedings of International Reliability Physics Symposium, IEEE, 1996, pp. 188–194.
[16] C.M. Tan, B.K.E. Chen, G. Xu, Y. Liu, Analysis of humidity effects on the degradation of high-power white LEDs, Microelectronics Reliability 49 (9–11) (2009) 1226–1230.
[17] M.G. Pecht, M.H. Chang, Failure mechanisms and reliability issues in LEDs, Solid State Lighting Reliability (2013) 43–110.
[18] L. Guoguang, H. Yun, E. Yunfei, Y. Shaohua, L. Zhifeng, The relationship between LED package and reliability, in: 2009 16th IEEE International Symposium on the Physical and Failure Analysis of Integrated Circuits, IEEE, July 2009, pp. 323–326.
[19] F. Wu, W. Zhao, S. Yang, C. Zhang, Failure modes and failure analysis of white LEDs, in: 2009 9th International Conference on Electronic Measurement & Instruments, IEEE, August 2009, pp. 4–978.
[20] S.C. Yang, P. Lin, C.P. Wang, S.B. Huang, C.L. Chen, P.F. Chiang, M.T. Chu, Failure and degradation mechanisms of high-power white light emitting diodes, Microelectronics Reliability 50 (7) (2010) 959–964.
[21] Z. Chen, Q. Zhang, F. Jiao, R. Chen, K. Wang, M. Chen, S. Liu, Study on the reliability of application-specific LED package by thermal shock testing, failure analysis, and fluid–solid coupling thermo-mechanical simulation, IEEE Transactions on Components, Packaging and Manufacturing Technology 2 (7) (2012) 1135–1142.
[22] G. Elger, D. Müller, A. Hanß, M. Schmid, E. Liu, U. Karbowski, R. Derix, Transient thermal analysis for accelerated reliability testing of LEDs, Microelectronics Reliability 64 (2016) 605–609.
[23] A. Hanss, E. Liu, M. Schmid, D. Müller, U. Karbowski, R. Derix, G. Elger, New method to separate failure modes by transient thermal analysis of high power LEDs, in: 2017 IEEE 67th Electronic Components and Technology Conference (ECTC), IEEE, May 2017, pp. 1136–1144.
[24] L. Guoguang, Y. Shaohua, L. Zhifeng, Failure analysis of LEDs, in: 2011 International Symposium on Advanced Packaging Materials (APM), IEEE, October 2011, pp. 69–72.
[25] L. Zhang, Y. Zhu, H. Chen, K. Leung, Y. Wu, J. Wu, Failure analysis on reflector blackening between lead frame electrodes in LEDs under WHTOL test, Microelectronics Reliability 55 (5) (2015) 799–806.
[26] G. Zhang, S. Feng, Z. Zhou, J. Liu, J. Li, H. Zhu, Thermal fatigue characteristics of die attach materials for packaged high-brightness LEDs, IEEE Transactions on Components, Packaging and Manufacturing Technology 2 (8) (2012) 1346–1350.
[27] B.H. Cumpston, I.D. Parker, K.F. Jensen, In situ characterization of the oxidative degradation of a polymeric light emitting device, Journal of Applied Physics 81 (8) (1997) 3716–3720.
[28] P.R. Griffiths, J.A. De Haseth, Fourier Transform Infrared Spectrometry, vol 171, John Wiley & Sons, 2007.

[29] A. Nyczyk, C. Paluszkiewicz, M. Hasik, M. Cypryk, P. Pospiech, Cross-linking of linear vinylpolysiloxanes by hydrosilylation—FTIR spectroscopic studies, Vibrational Spectroscopy 59 (2012) 1—8.
[30] J. Kim, Non-destructive characterization of railway materials and components with infrared thermography technique, Materials 12 (24) (2019) 4077.
[31] H. Glavaš, T. Barić, D. Jukić, E. Desnica, Infrared thermography as evaluation method of a faulty status of a LED LCD TV, Journal of the Society for Information Display 28 (12) (2020) 936—955.
[32] M. Meneghini, C. De Santi, M. Buffolo, A. Munaretto, G. Meneghesso, E. Zanoni, Towards high reliability GaN LEDs: understanding the physical origin of gradual and catastrophic failure, in: 2015 12th China International Forum on Solid State Lighting (SSLCHINA), IEEE, November 2015, pp. 63—66.
[33] R.K. Kasim, Y. Cheng, M. Pomerantz, R.L. Elsenbaumer, Investigation of device failure mechanisms in polymer light-emitting diodes, Synthetic Metals 85 (1—3) (1997) 1213—1214.
[34] H.H. Yen, H.C. Kuo, W.Y. Yeh, Particular failure mechanism of GaN-based alternating current light-emitting diode induced by GaO $ _ {x} $ oxidation, IEEE Photonics Technology Letters 22 (15) (2010) 1168—1170.
[35] Y. Cao, N. Wang, H. Tian, J. Guo, Y. Wei, H. Chen, W. Huang, Perovskite light-emitting diodes based on spontaneously formed submicrometre-scale structures, Nature 562 (7726) (2018) 249—253.
[36] K. Pantzas, G. Patriarche, D. Troadec, S. Gautier, T. Moudakir, S. Suresh, A. Ougazzaden, Nanometer-scale, quantitative composition mappings of InGaN layers from a combination of scanning transmission electron microscopy and energy dispersive x-ray spectroscopy, Nanotechnology 23 (45) (2012) 455707.
[37] C.M. Tan, B.E. Chen, Y.Y. Foo, R.Y. Chan, G. Xu, Y.J. Liu, Humidity effect on the degradation of packaged ultra-bright white LEDs, in: 2008 10th Electronics Packaging Technology Conference, IEEE, December 2008, pp. 923—928.
[38] J. Xu, M.F. Schubert, A.N. Noemaun, D. Zhu, J.K. Kim, E.F. Schubert, Y. Park, Reduction in efficiency droop, forward voltage, ideality factor, and wavelength shift in polarization-matched GaInN/GaInN multi-quantum-well light-emitting diodes, Applied Physics Letters 94 (1) (2009) 011113.
[39] D. Zhu, J. Xu, A.N. Noemaun, J.K. Kim, E.F. Schubert, M.H. Crawford, D.D. Koleske, The origin of the high diode-ideality factors in GaInN/GaN multiple quantum well light-emitting diodes, Applied Physics Letters 94 (8) (2009) 081113.
[40] A. David, C.A. Hurni, N.G. Young, M.D. Craven, Electrical properties of III-Nitride LEDs: recombination-based injection model and theoretical limits to electrical efficiency and electroluminescent cooling, Applied Physics Letters 109 (8) (2016) 083501.
[41] X. Guo, E.F. Schubert, Current crowding and optical saturation effects in GaInN/GaN light-emitting diodes grown on insulating substrates, Applied Physics Letters 78 (21) (2001) 3337—3339.
[42] Z.A. Gilani, M.S. Awan, I. Ahmad, Y. Tan, Characterization of InGaN by means of I—V measurements of respective light-emitting diode (LED) by DLTS, Arabian Journal for Science and Engineering 40 (1) (2015) 263—268.
[43] C.M. Tan, P. Singh, Time evolution degradation physics in high power white LEDs under high temperature-humidity conditions, IEEE Transactions on Device and Materials Reliability 14 (2) (2014) 742—750.
[44] P. Singh, C.M. Tan, L.B. Chang, Early degradation of high power packaged LEDs under humid conditions and its recovery—myth of reliability rejuvenation, Microelectronics Reliability 61 (2016) 145—153.
[45] Z. Ma, H. Cao, S. Lin, X. Li, L. Zhao, Degradation and failure mechanism of AlGaN-based UVC-LEDs, Solid-State Electronics 156 (2019) 92—96.

[46] A. Jayawardena, N. Narendran, Analysis of electrical parameters of InGaN-based LED packages with aging, Microelectronics Reliability 66 (2016) 22–31.
[47] J. Mertz, Introduction to Optical Microscopy, Cambridge University Press, 2019.
[48] P. Török, F.J. Kao (Eds.), Optical Imaging and Microscopy: Techniques and Advanced Systems, vol 87, Springer, 2007.
[49] T.F. Watson, Applications of confocal scanning optical microscopy to dentistry, British Dental Journal 171 (9) (1991) 287–291.
[50] A. Clarke, C. Eberhardt, C.N. Eberhardt, Microscopy Techniques for Materials Science, Woodhead Publishing, 2002.
[51] Y. Leterrier, L. Medico, F. Demarco, J.A. Månson, U. Betz, M.F. Escola, F. Atamny, Mechanical integrity of transparent conductive oxide films for flexible polymer-based displays, Thin Solid Films 460 (1–2) (2004) 156–166.
[52] H. Kim, H. Yang, C. Huh, S.W. Kim, S.J. Park, H. Hwang, Electromigration-induced failure of GaN multi-quantum well light emitting diode, Electronics Letters 36 (10) (2000) 908–910.
[53] J. Park, Y.G. Sin, J.H. Kim, J. Kim, Dependence of adhesion strength between GaN LEDs and sapphire substrate on power density of UV laser irradiation, Applied Surface Science 384 (2016) 353–359.
[54] P. Singh, C.M. Tan, Degradation physics of high power LEDs in outdoor environment and the role of phosphor in the degradation process, Scientific Reports 6 (1) (2016) 1–13.
[55] P. Singh, C.M. Tan, Uncover the degradation science of silicone under the combined temperature and humidity conditions, IEEE Access 6 (2017) 1302–1311.
[56] J. Schmitt, H.C. Flemming, FTIR-spectroscopy in microbial and material analysis, International Biodeterioration & Biodegradation 41 (1) (1998) 1–11.
[57] C. Berthomieu, R. Hienerwadel, Fourier transform infrared (FTIR) spectroscopy, Photosynthesis Research 101 (2) (2009) 157–170.
[58] M.J. Pelletier (Ed.), Analytical Applications of Raman Spectroscopy, vol 427, Blackwell science, Oxford, 1999.
[59] L.A. Lyon, C.D. Keating, A.P. Fox, B.E. Baker, L. He, S.R. Nicewarner, M.J. Natan, Raman spectroscopy, Analytical Chemistry 70 (12) (1998) 341–362.
[60] S. Watzke, P. Altieri-Weimar, Degradation of silicone in white LEDs during device operation: a finite element approach to product reliability prediction, in: 2014 15th International Conference on Thermal, Mechanical and Mulit-Physics Simulation and Experiments in Microelectronics and Microsystems (EuroSimE), IEEE, April 2014, pp. 1–5.
[61] R.J. Bartholomew, Raman Spectral Studies of Solutions of Formic Acid and Methyl Formate, 1996.
[62] K. Wang, F. Chen, Z. Liu, X. Luo, S. Liu, Design of compact freeform lens for application specific light-emitting diode packaging, Optics Express 18 (2) (2010) 413–425.
[63] Z. Mao, X. Luo, J. Yang, S. Liu, Moisture diffusivity analysis of polycarbonate for led lens, in: 2010 11th International Conference on Electronic Packaging Technology & High Density Packaging, IEEE, August 2010, pp. 1080–1087.
[64] M.Y. Mehr, M.R. Toroghinejad, F. Karimzadeh, W.D. Van Driel, G.Q. Zhang, A review on discoloration and high accelerated testing of optical materials in LED based-products, Microelectronics Reliability 81 (2018) 136–142.
[65] A. Rivaton, B. Mailhot, J. Soulestin, H. Varghese, J.L. Gardette, Comparison of the photochemical and thermal degradation of bisphenol-A polycarbonate and trimethylcyclohexane–polycarbonate, Polymer Degradation and Stability 75 (1) (2002) 17–33.

[66] G. Lu, M.Y. Mehr, W.D. Van Driel, X. Fan, J. Fan, K.M.B. Jansen, G.Q. Zhang, Color shift investigations for LED secondary optical designs: comparison between BPA-PC and PMMA, Optical Materials 45 (2015) 37–41.
[67] G. Lu, W.D. Van Driel, X. Fan, M.Y. Mehr, J. Fan, C. Qian, G.Q. Zhang, Colour shift and mechanism investigation on the PMMA diffuser used in LED-based luminaires, Optical Materials 54 (2016) 282–287.
[68] J. Li, Y. Tang, Z. Li, X. Ding, L. Rao, B. Yu, Investigation of stability and optical performance of quantum-dot-based LEDs with methyl-terminated-PDMS-based liquid-type packaging structure, Optics Letters 44 (1) (2019) 90–93.
[69] L. Suslik, D. Pudis, M. Goraus, R. Nolte, J. Kovac, J. Durisova, P. Schaaf, Photonic crystal and photonic quasicrystal patterned in PDMS surfaces and their effect on LED radiation properties, Applied Surface Science 395 (2017) 220–225.
[70] P. Singh, C.M. Tan, Time evolution of packaged LED lamp degradation in outdoor applications, Optical Materials 86 (2018) 148–154.
[71] D. Sayre, H.N. Chapman, X-ray microscopy, Acta Crystallographica Section A: Foundations of Crystallography 51 (3) (1995) 237–252.
[72] R. Chen, Q. Zhang, T. Peng, F. Jiao, S. Liu, Failure analysis techniques for high power light emitting diodes, in: 2011 12th International Conference on Electronic Packaging Technology and High Density Packaging, IEEE, August 2011, pp. 1–4.
[73] https://globalledoled.com/articles_&_papers/advanced-packaging-x-ray-inspection-and-the-humble-led/.
[74] X.P. Maldague, Introduction to NDT by active infrared thermography, Materials Evaluation 60 (9) (2002) 1060–1073.
[75] C. Meola, G.M. Carlomagno, Recent advances in the use of infrared thermography, Measurement Science and Technology 15 (9) (2004) R27.
[76] C.M. Tan, P. Singh, W. Zhao, H.C. Kuo, Physical limitations of phosphor layer thickness and concentration for white LEDs, Scientific Reports 8 (1) (2018) 1–14.
[77] K.C.A. Smith, C.W. Oatley, The scanning electron microscope and its fields of application, British Journal of Applied Physics 6 (11) (1955) 391.
[78] T.L. Hayes, Scanning electron microscope techniques in biology, in: Advanced Techniques in Biological Electron Microscopy, Springer, Berlin, Heidelberg, 1973, pp. 153–214.
[79] S. Zaefferer, S.I. Wright, D. Raabe, Three-dimensional orientation microscopy in a focused ion beam–scanning electron microscope: a new dimension of microstructure characterization, Metallurgical and Materials Transactions A 39 (2) (2008) 374–389.
[80] F. Wu, Y. Wu, B. An, F. Wu, Analysis of dark stain on chip surface of high-power LED, in: 2006 7th International Conference on Electronic Packaging Technology, IEEE, August 2006, pp. 1–4.
[81] X. Luo, J. Fan, M. Zhang, C. Qian, X. Fan, G. Zhang, Degradation mechanism analysis for phosphor/silicone composites aged under high temperature and high humidity condition, in: 2017 18th International Conference on Electronic Packaging Technology (ICEPT), IEEE, August 2017, pp. 1331–1336.
[82] Z. Yu, S. Boseck, Scanning acoustic microscopy and its applications to material characterization, Reviews of Modern Physics 67 (4) (1995) 863.
[83] B.T. Khuri-Yakub, Scanning acoustic microscopy, Ultrasonics 31 (5) (1993) 361–372.
[84] R.J. da Fonseca, L. Ferdj-Allah, G. Despaux, A. Boudour, L. Robert, J. Attal, Scanning Acoustic Microscopy—recent applications in materials science, Advanced Materials 5 (7-8) (1993) 508–519.
[85] C. Zhaohui, Z. Qin, W. Kai, L. Xiaobing, L. Sheng, Reliability test and failure analysis of high power LED packages, Journal of Semiconductors 32 (1) (2011) 014007.

[86] C.P. Liu, Y.F. Liu, C.H. Li, H.C. Cheng, Y.C. Kung, J.Y. Lin, A novel decapsulation technique for failure analysis of epoxy molded IC packages with Cu wire bonds, Microelectronics Reliability 52 (4) (2012) 725–734.
[87] Petr, N., Milan, A., & Petr, S. Component Package Decapsulation Process with Analogue Signature Analysis Support.
[88] S. Manoharan, C. Patel, P. McCluskey, Decapsulation of plastic encapsulated microelectronics with copper wire bonds, Additional Papers and Presentations 2016 (HiTEC) (2016) 000092–000096.

## Further reading

[1] H.H. Shin, J.H. Kim, B.Y. Han, J.S. Yoo, Failure analysis of a phosphor-converted white light-emitting diode due to the CaS: Eu phosphor, Japanese Journal of Applied Physics 47 (5R) (2008) 3524.
[2] M.Y. Mehr, W.D. Van Driel, K.M.B. Jansen, P. Deeben, M. Boutelje, G.Q. Zhang, Photodegradation of bisphenol A polycarbonate under blue light radiation and its effect on optical properties, Optical Materials 35 (3) (2013) 504–508.

CHAPTER 3

# Effect of environment on white packaged LED reliability

## 3.1 Introduction

Because of their longer life span, reliability, and effectiveness, LED applications have grown in popularity over the last two decades [1—3]. Cree can now produce 303 lumens per watt from a white high-power LED at room temperature, which is a 10% improvement over their previous record of 276 lumens per watt (until 2021) [4]. While the increase in lumen efficacy over time is remarkable, their reliability investigation is still lacking, especially for LEDs used in outdoor applications where temperatures and humidity are higher. As shown in Fig. 1.2 in Chapter 1, their outdoor applications account for more than 35% of total applications (until 2019).

Thermal reliability has been studied a lot more than humidity reliability in the reliability analysis of high-power LEDs, as shown in Fig. 1.4 in Chapter 1, but humidity reliability studies are on the rise [5]. Moisture-related failures, on the other hand, are equally significant, particularly in outdoor applications. In this chapter, effect of various stress factors such as temperature, humidity, and electrical will be discussed.

### 3.1.1 High-temperature assessment

LEDs have a longer life span than other light sources, which is one of their advantages [6,7]. When an LED's thermal architecture is ineffective for heat dissipation, resulting in high junction temperature during LED activity, the LED's operating life can be much shorter than its theoretical value [8—11]. LEDs are also experiencing temperature cycling during their surface mount process, and delamination can occur if the process parameters setting are not properly set. Such delamination renders weak mechanical reliability for the LED packages [12—14]. Therefore, for good optical efficiency, thermal and mechanical reliability, LED package design using appropriate materials with good structure design are essential [15]. With increasing higher power for

*Reliability and Failure Analysis of High-Power LED Packaging*
ISBN 978-0-12-822408-3
https://doi.org/10.1016/B978-0-12-822408-3.00002-2

the LEDs used today, degradation issues due to thermally induced stresses as resulted from the different coefficient of thermal expansion (CTE) of the LED die and its mounting substrate should be addressed [16].

## 3.2 Thermal stress in LED chips

Mismatch of the CTE of different layers in LED die can also render unwanted residual stress occurring in LEDs after the metal-organic chemical vapor deposition (MOCVD) process [17].

In other words, residue stresses can occur during the fabrication, packaging, and operations of LED chips, and these stresses can cause wafer bending, dislocation forming, or cracking [18].

Although sapphire substrate is commonly used for GaN-based LED die, the large lattice mismatch and process-induced thermoelastic effects owing to the different CTE of the sapphire substrate and the GaN die produce high compressive stress in the die [19,20]. This compressive stress can cause distortion in the band structure of GaN, affecting its efficiency of the electron—hole recombination, threshold current for light emission, and the peak wavelength of the light emission [21]. The poor thermal conductivity of sapphire as substrate is also considered as the primary blockage to the use of high-power LEDs. Silicon substrate with much better thermal conductivity has been used to demonstrate the possibility of increasing the power of the LEDs [21,22]. However, the huge mismatch in the lattices between GaN and Si results in the formation of dislocations and a highly stressed GaN layer, reducing the internal efficiency of light generation.

To overcome this lattice match, laser lift-off (LLO) and wafer bonding process are two methods employed to have a thermally conductive substrate to replace the sapphire substrate [23]. In LLO method, the poor thermal conductivity sapphire substrate is life-off using laser LLO after the GaN epilayer is bonded to a good thermal conductive Si substrate via a metallization joint [21]. The metallization used is gold in general. It is found that the thickness of the Au bonding has a clear impact on stress relief in GaN as shown in Fig. 3.1.

Chou et al. [24] demonstrated the wafer bonding method to replace the GaAs substrate by good thermal conductive substrates as shown in Fig. 3.2, and improvement in the light extraction efficiency and thermal performance of AlGaInP LEDs was observed. Unfortunately, it was found that the stress built up either during the GaAs substrate removal etching process or the subsequent annealing process can cause device crack. Through

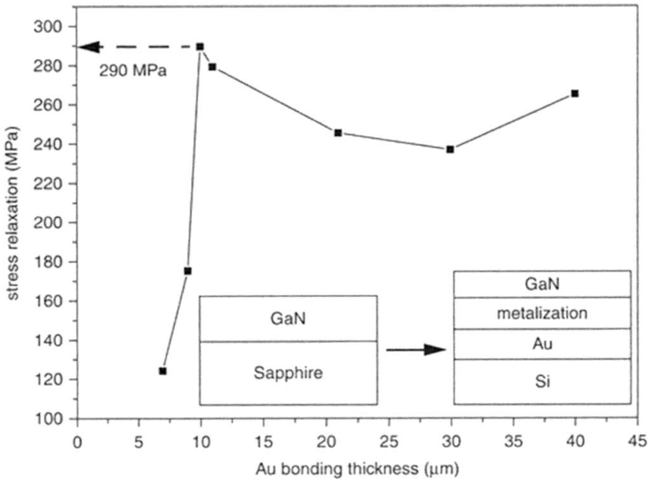


**Figure 3.1** The tension relief versus the thickness of the Au bonding sheet [21]. (Applied Physics Letters. *Reprinted with permission.)*

optimizing the properties and dimensions of the epilayer and the thermal conductive substrate, they found that the buildup stress can be reduced with enhanced LED reliability as shown in Fig. 3.3.

For better thermal management of high-power LEDs, Tang et al. created an integrated heat sink with vapor chamber (IHSVC) [25]. As shown in Fig. 3.4, the wick, which is a key component of the IHSVC, was constructed with parallel and orthogonal microgrooves. These microgrooves were created using the micromilling process, and tests were carried out to see if the package's optical and thermal performance changed as a result of these changes. In their study, the temperature distribution of the system was found to be very uniform, and a negative effect of junction temperature on radiant efficacy was discovered, resulting in an improvement in the system's overall luminous efficacy.

## 3.3 Thermal stress in die attachment

Die attachment is used to attach LED chip to a substrate or heat sink for their mechanical integrity and heat dissipation. Upon cooling of the package after solder reflow or polymer curing, or other temperature adjusting methods, the different CTE of the materials in multilayered structure in LED packages results in thermal stress within the die attachment and the chips, and issues such as cracks and delamination appear. With the

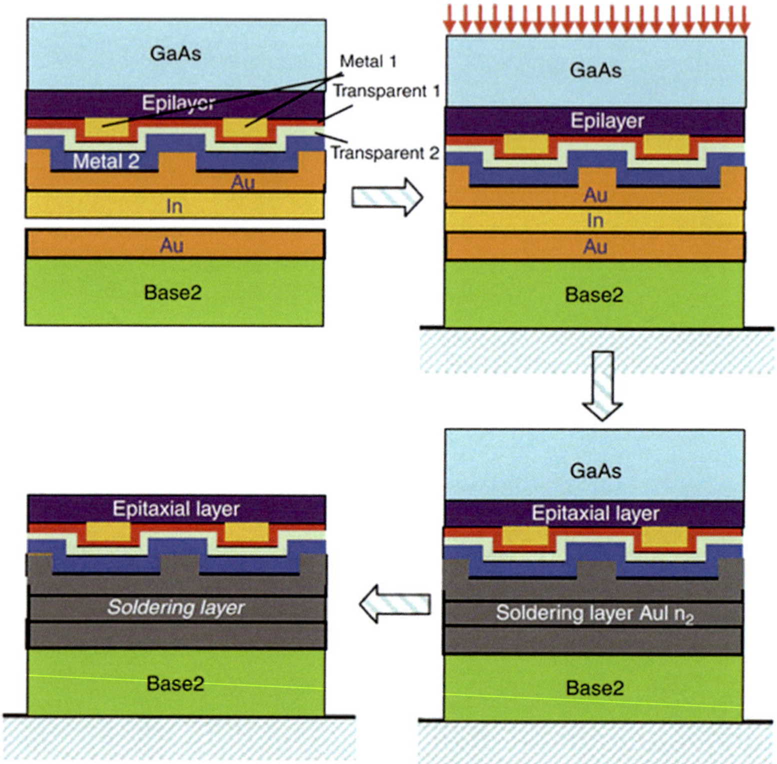


**Figure 3.2** Fabrication steps of the high brightness AlGaInP LEDs [24]. *(Microelectronics Reliability, Elsevier. Reprinted with permission.)*

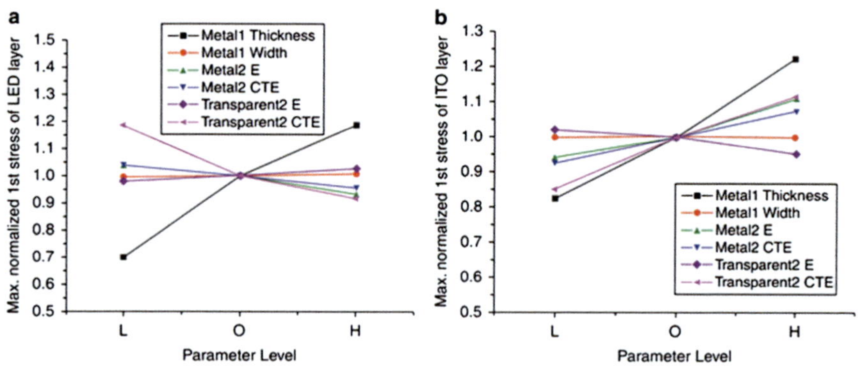


**Figure 3.3** The epitaxial layer (a) and the first transparent layer (b) have their maximum first stress variance normalized [24]. *(Microelectronics Reliability, Elsevier. Reprinted with permission.)*

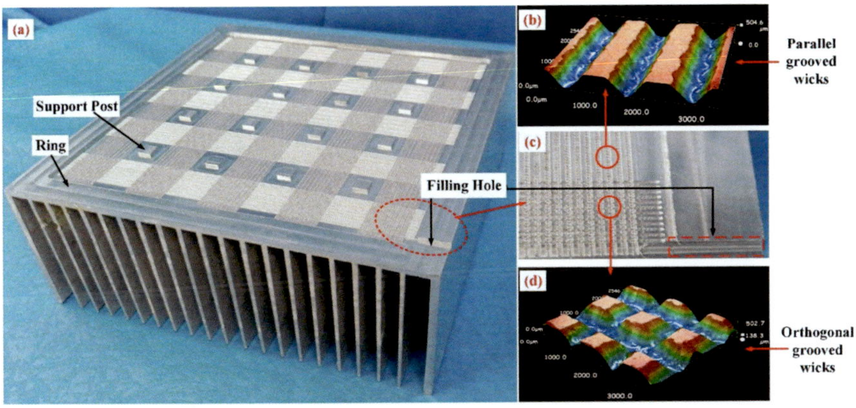


**Figure 3.4** The IHSVC's condenser and a scanned 3D view of the wicks [25]. *(*Physical review B. *Reprinted with permission.)*

LED packages further exposed to power cycling, thermal cycling, or thermal shock, they will fail early, resulting in reliability issues. Therefore, choosing materials with closely matched CTE in the LED package design is important. The thermal and mechanical properties of materials widely used in LED packages are shown in Table 3.1.

In LED packages, die attachment is a key element in the junction to ambient heat conduction path [12]. The delamination between the die attachment and chip mentioned above will block the heat dissipation and degrades LED output via increasing its junction temperature. Fig. 3.5 shows the thermal resistances of the samples without and with 3, 6, and 24 h of moisture preprocessing. Their corresponding thermal resistances are found to be 41.7 K/W, 52.2 K/W, 52.2 K/W, and 70.9 K/W, respectively.

**Table 3.1** LED packaging materials' thermal and mechanical properties [26].

| Material | SiC | Die attach | Lead frame | Mold | Epoxy lens |
|---|---|---|---|---|---|
| $k$ (W/M °C) | 200 | 7.5 | 401 | 0.23 | 0.17 |
| $C$(J/kg °C) | 1000 | 300 | 385 | 1256 | 1173 |
| $\rho$(kg/m$^3$) | 2975 | 2400 | 8300 | 1300 | 980 |
| $\alpha$, CTE($10^{-6}$/K) | 14 | 30 | 18 | 45 | 45 |
| Poison's ratio | 0.31 | 0.35 | 0.34 | 0.35 | 0.49 |
| $E$ (GPa) | 110 | 40 | 190 | 5.2 | 0.5 |

Elsevier. Reprinted with permission.

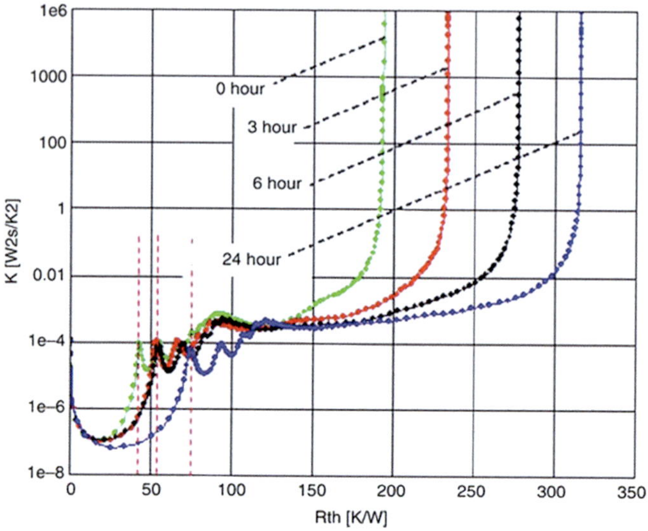


**Figure 3.5** Differential structure functions of LED packages with different severity of delamination [12]. (Journal of Crystal Growth. *Reprinted with permission.)*

Therefore, we can see that the integrity of the die attachment bonding is crucial to the durability of LEDs.

Another significant material property that influences thermal stress is elastic modulus. It is a measurement of a material's flexibility. When the elastic modulus of a material is high, it has a low thermal stress when the temperature changes.

For polymer-based DA, the glass transition temperature (Tg) [27] should also be considered. At Tg, the CTE, heat power, and a few other properties of a polymer-based DA change abruptly. Above Tg, CTE typically increases dramatically. As a result, it's important that packaged LEDs operate at temperatures lower than Tg.

The curing profile is another important aspect to take note. Both the shrinkage of the adhesive during curing and the polymerization during cooling from the curing temperature can cause residual stresses [28]. How an adhesive is cured and the cure temperature/time are significant to its thermomechanical properties and stresses. Phase curing or allowing the adhesive to gel at a temperature lower than the final cure temperature was found to reduce stresses dramatically [28].

Thermal diffusivity is a parameter that signifies how quickly energy can be transferred into a solid material. High thermal diffusivity indicates that heat energy can disperse quickly into the material, resulting in lower thermal stress during transient temperature changes.

Flexible, elastomeric resin-based adhesives are available commercially, and they can be used to relax stresses. They have low elastic modulus and can be used as low-stress or "no-stress" adhesives [29]. A number of recommendations can be followed to relieve stress [29]:

- Select an adhesive with a low shrinkage during cure (less than 3%).
- Choose an adhesive that fits the adherends' CTEs closely after curing
- Select an adhesive with a low modulus of elasticity
- Modifying cure conditions, such as phase cure rather than quick high-temperature shock cure.
- Select an adhesive with higher glass transition temperature than the expected operation or test temperature

While the above recommendations are useful, they cannot ensure no voids formation in the die attachment [30] and increase in thermal resistance of the LED package can still exist, which in turn causes localized stress concentration on the chip with possible die cracking during the bonding phase or thermal cycling. Such voids are created during die attachment where gas bubbles could be trapped during the attachment process.

Most die crack failures can be traced to the poor die attachment. Consequently, thoroughly washing the surfaces before applying die attachment (DA), degassing before applying DA, proper selection of DA material with good wettability to the surfaces, performing the bonding process in a clean setting, etc., must be practiced to remove the voids created in the DA process.

## 3.4 Thermal stress in LED packaging

As shown in Fig. 3.6, reliability problems caused by thermal stress can occur in a variety of places in LED packages, including the wire bond, LED lens, PCB, and so on.

Sim et al. proposed a low-temperature cofired ceramic-chip-on-board (LTCC-COB) kit with improved thermal properties [31]. Fig. 3.6b depicts a diagram of the architecture. There was no insulation layer between the metal base and the LED chip in this design. When compared to traditional surface mount system PCB-based packaged LED designs, the design produced higher performance. The proposed design's EL peak intensity was found to be 1.75 times higher than that of an Surface Mount device, SMD-PCB based LED kit. Due to its excellent properties, this

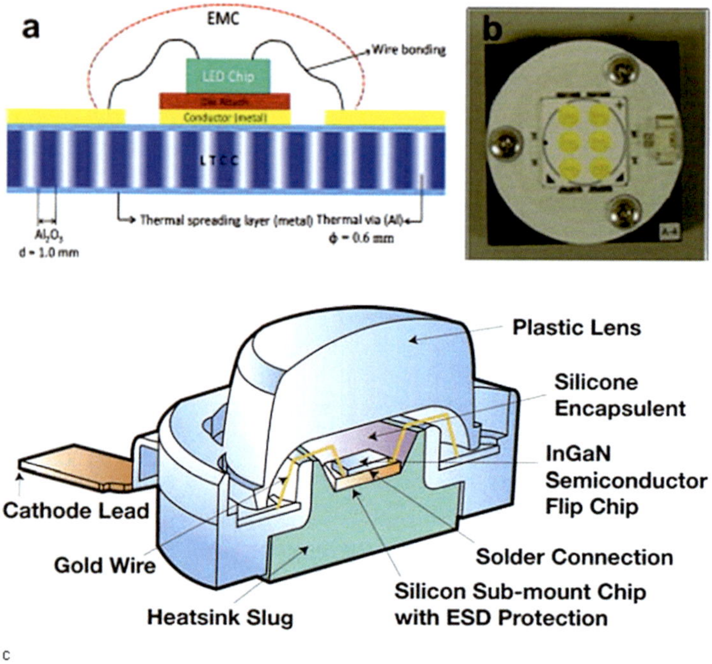


**Figure 3.6** (a) LTCC-COB package simplified model. (b) LTCC-COB kit LED lamp assembly [31] photo. (c) A typical LED kit diagram. *(Elsevier. Reprinted with permission.)*

structure may be a promising candidate for illumination modules and high-power LED package design.

LED chips are encapsulated with silicone lens for their environmental protection and color tuning. These encapsulants are injected at high temperature while they are in liquid form and solidify thereafter. Again, due to the differential thermal expansivities of the materials, residual stresses for the microinjection-molded lens could be concentrated in the regions where drastic geometrical changes occur, according to Chen Weng et al. [32], and molding temperature becomes an important processing parameter that will affect these residual stresses.

Another thermal-stress-related wear-out mechanism in LED packaging is the thermomechanical stress-induced wire ball bond fatigue [33]. Repeated cycles with wide temperature range lead to rapid degradation. With the expansion of the encapsulant force, the wire bond moves away from the die's surface [34–36], which will be returned upon cooling down as shown in Fig. 3.7. Prolonged and repeated thermal expansion and contraction can result in the wire bond fatigue. Such fatigue is affected by

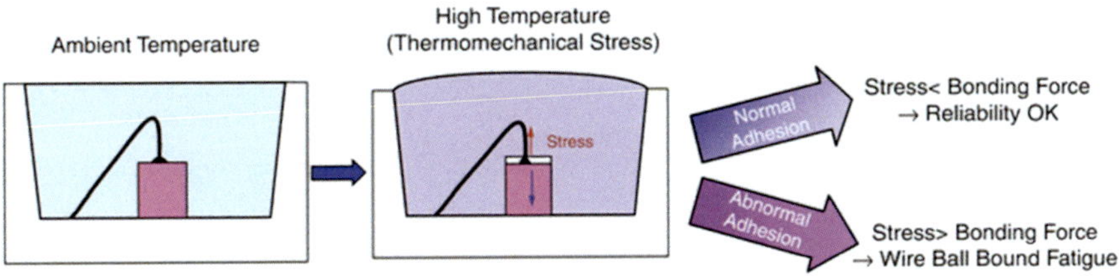


**Figure 3.7** Bond fatigue in wire balls [33]. *(Elsevier. Reprinted with permission.)*

the CTE, Young's modulus of the encapsulant, and the hardness of the die at different operating temperatures. The length of the bond wire and its loop height can also affect the bonding joint's reliability [37].

Besides the CTE mismatch that causes thermomechanical stresses as discussed earlier, elastic modulus of the materials is also another factor to be considered. For example, Hu et al. [15] used ceramic mold instead of plastic because of the closer matched CTE of the mold and the chip, the thermomechanical stress level in the mold was found to be higher than the case with plastic mold as can be seen in their comparison in Figs. 3.8 and 3.9.

## 3.5 Temperature cycling effect on LEDs

Ball grid array (BGA) packages are very common in LED technology today, and Vries et al. examined their durability behavior under a thermal shock test and a thermal cycle test [38]. Fig. 3.10 depicts the profiles for the two measures.

Finite element analyses and experiments on BGA packages that were subjected to the two forms of thermal stress tests were performed in their study. The main distinction between the two experiments is the temperature ramp speeds. BGA packages with 256 solder balls in four peripheral rows were soldered onto a four-layer panel with a thickness of 1.6 mm for this experiment. The packages were dried at 125 °C before being assembled. A 150 μm thick stencil was used to print solder paste onto the plates. The assemblies were soldered in a reflow furnace with a fixed peak temperature of 260 °C. Fig. 3.11 shows one-quarter of the finite element model of a single BGA fixed on a printed circuit board. Surface evolver was used to determine the precise shape of the vital solder ball. Adjusted linear elastic beam elements were used to model the other solder balls. The properties of the beams were tweaked to match the solder joint's connect. A two-solder ball model was used to evaluate the creep and platicity behavior. Creep and plasticity properties are present in the essential solder ball.

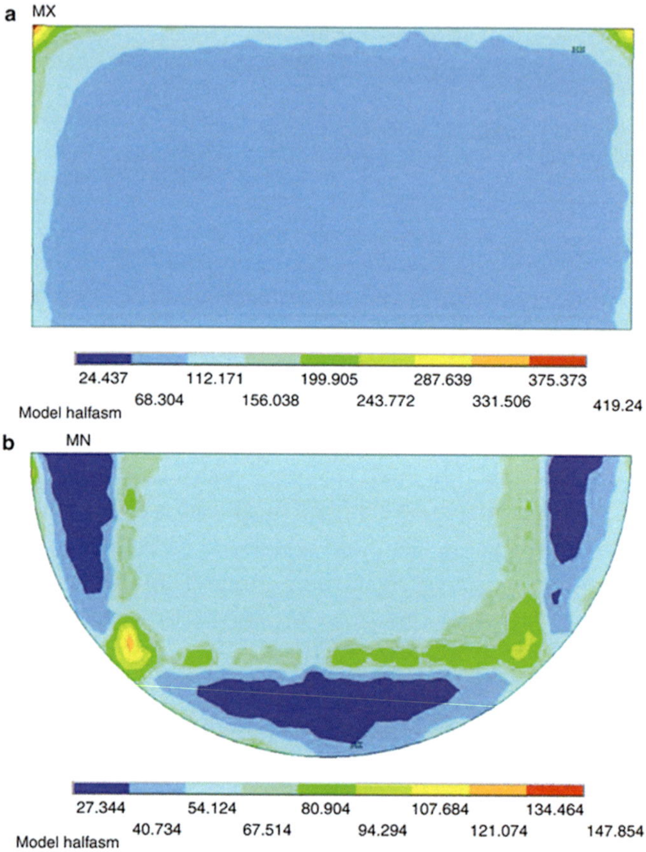


**Figure 3.8** Thermomechanical stress distribution in LEDs with ceramic packages as computed using finite element analysis. (a) The chip itself; (b) the encapsulant [15].

On the failed samples, cross sections were taken. Some standard examples are shown in Fig. 3.12. Cracks can be seen in the solder that run close to the $Cu_6Sn_5$-intermetallic layer on the component's side. The failure mode is the same in both experiments, both on the appearance of the deformation and the position of the cracks, which are often run through the solder.

Fig. 3.13 shows the averaged cumulative inelastic strain in solder ball in one step of thermal shock test and the thermal cycle test for isothermal calculations. Therefore, thermal shock experiments appear to be beneficial since they greatly accelerate failures. However, cautious is needed in performing the tests.

Stresses in temperature cycle tests are primarily caused by the mismatches in thermal expansion coefficients of the materials. On the other

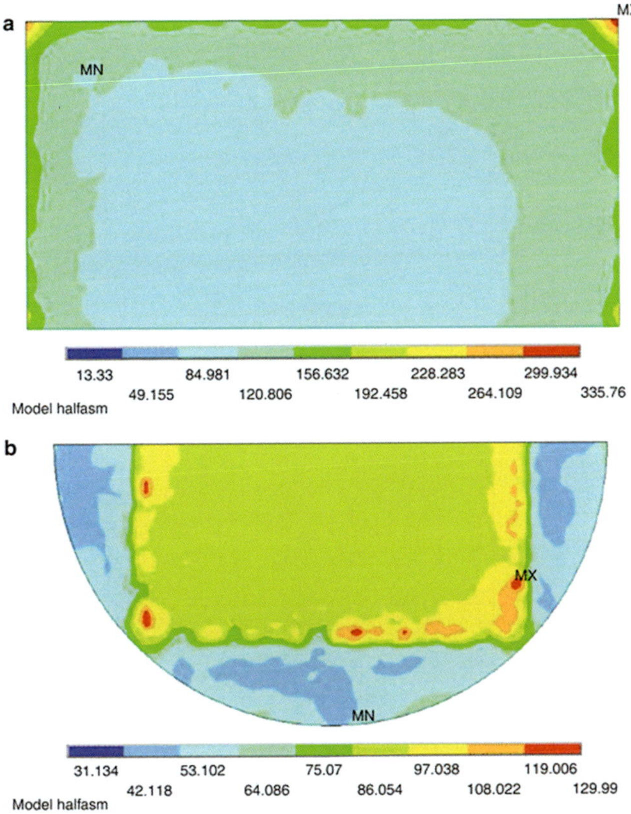


**Figure 3.9** Thermomechanical stress distribution in LEDs with plastic packages as estimated using finite element analysis. (a) The chip itself; (b) the encapsulant [15].

hand, additional stresses can be produced in thermal shock tests due to temperature gradients. The temperature gradient effect is dependent on the thermal behavior of the tested device. Consequently, disparate test results can be obtained from the temperature cycle and thermal shock test. The thermal shock test can result in a variety of failures. Temperature cycling tests are generally used to test the resisting nature of parts under the cyclical exposures of extreme temperatures. Similarly thermal shock tests also accelerate fatigue failures, which resemble cycling tests.

From the abovementioned investigations, the following can be summarized:

**(1)** After a thermal shock test, the optical and electronic output of an LED package degrades more than after a thermal cycling test. During the

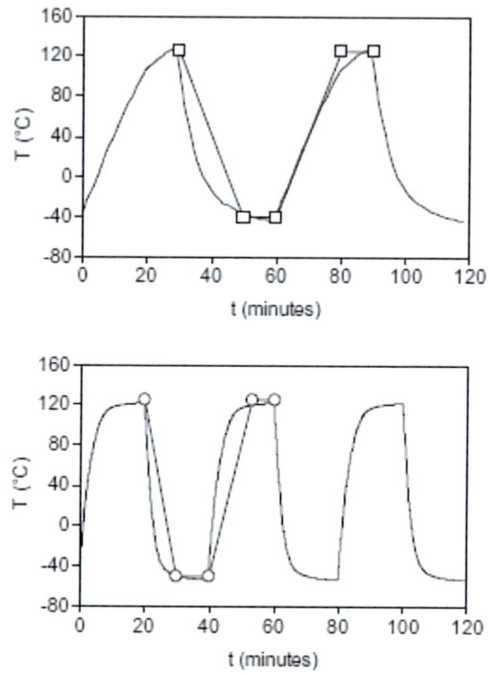


**Figure 3.10** Temperature profile of temperature cycle (top) and thermal shock (bottom) tests [38]. *(Elsevier. Reprinted with permission.)*

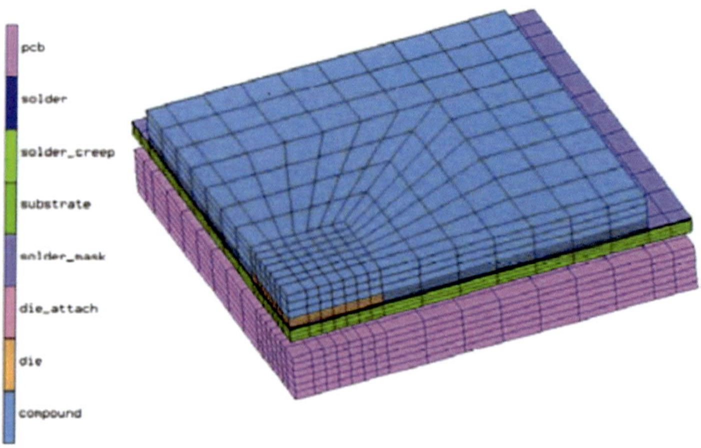


**Figure 3.11** The finite element model's geometry [38]. *(Elsevier. Reprinted with permission.)*

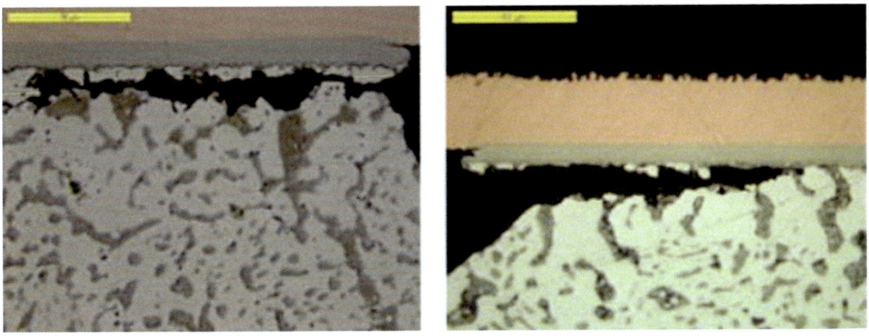

**Figure 3.12** Failures that are common after an examination. Fatigue cracks run along the component's edge. Cycle test on the left, shock test on the right [38]. *(Elsevier. Reprinted with permission.)*

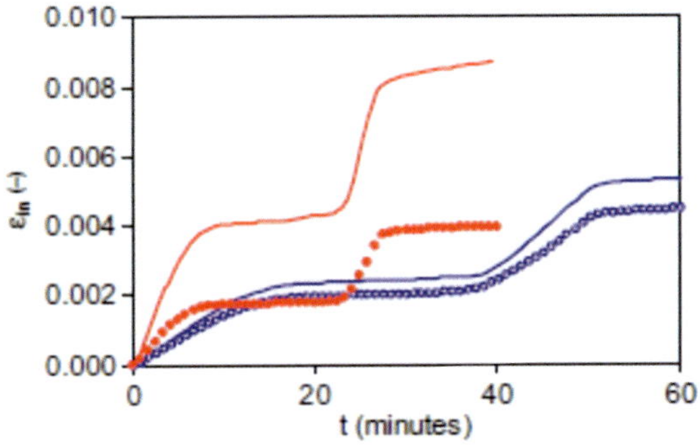


**Figure 3.13** Inelastic strain accumulated in a solder ball as a function of time. The blue dots show the results of isothermal temperature cycle test, red dots show the results of isothermal thermal shock test, blue line show the results of semitransient temperature cycle test, and red line show the results of semitransient thermal shock test [38]. *(Elsevier. Reprinted with permission.)*

thermal shock test, some LED package samples showed catastrophic failure.

**(2)** The thermomechanical responses under thermal cycling and thermodynamics can be effectively simulated using sequential coupling thermal and mechanical modeling.

**(3)** The temperature gradient in the LED package during thermal cycling and thermal shock testing contributes significantly to the LED package's local deformation. The stress behavior of the gold wire is significantly affected by the silicone's time and temperature-dependent material property. Under thermal shock loading, the stress concentration occurred at the second bonding of gold wire, resulting in catastrophic failure of the LED package.

**(4)** Experiments like these, in combination with numerical simulation and failure mechanism analysis, can help to improve the reliability of high-power LED packages.

## 3.6 Electrical effect on LEDs

Several authors have stated that electric currents reduce the light output capacity of GaN-based blue LED chips [39–43]. The light output power will be degraded even at very low current densities [44,45]. The increase of nonradiative recombination centers in the active layer was generally blamed for the failure [46–48]. Nonradiative recombination rates will rise due to a number of mechanisms, including: (1) defect migration to the active layer [45,49], (2) point defects related to Mg element diffusion along dislocations to the active sheet [50,51], (3) nitrogen vacancy formation in the active layer [44].

LEDs are often joule-heated when they are subjected to electric current . Two techniques are commonly used to reduce joule-heating effects as much as possible: one is to increase thermal conductivities of devices so that the system temperature is kept as close to room temperature as possible, and the other is to use pulsed currents instead of direct currents [46,52,53]. At room temperature, Shao et al. [42] used a 600 mA dc current to stress unpackaged LED chips. The maximum amount of time for aging was 12 h. They found that air cooling the samples for half an hour before each series of measurements did not eliminate the effect of joule heating on LED chips. For joule heat dissipation, Buso et al. [43] installed LED chips on the TO18 metallic box. At room temperature, electric currents were used to stress the assembly. The current densities range from 35 to 85 $A/cm^2$. The longest aging period was over 1000 h. At low measurement current levels, both groups of work reported a substantial decrease in optical power. The reverse and low-forward bias current components of these samples showed an degradation in I–V characteristics, which was due to an increase in parasitic leakage routes. However, in Buso's experiments (stress condition 35–85 $A/cm^2$, room temperature), the aging duration is about 100 times longer than in Shao's experiments (stress condition 60 $A/cm^2$, room temperature), and the difference is due to their heat dissipating capacity.

"Electrical Over Stress" (EOS)/Electrostatic discharge (ESD)" is a serious hazard to LEDs that could be caused by the same power source. When the LED driving current or voltage reaches the component's maximum rating values, an EOS occurs. EOS/ESD can take several forms: some are created during the LED assembly or testing phase, while others are generated by the power supply or originate in the environment induced by the electromagnetic field [54].

EOS/ESD is a dangerous hazard to solid-state light technology because it is in the light output direction. Furthermore, the harm it causes is not always apparent immediately. The affected LED can stop working days or even months after installation in certain cases. The time between the occurrence of an ESD and the inability to locate the fault may be considerable, with significant consequences as more luminaires can be manufactured and installed, raising the cost of warranty replacement. The energy given by the EOS/ESD pulses determines the time lapse.

Although there are several causes for an EOS/ESD, there is only one result: LED failure. EOS/ESD causes damage to the LED chip structure, causing it to fail sooner than anticipated.

Renso et al. [54] investigated the degradation mechanisms and electro-optical properties of light emitting diodes (LEDs) that had been subjected to electrical overstress. They looked at the impact of different parameters on current crowding impacts, with EOS being one of them. Higher electrical stress causes more current crowding, which causes the higher current density, as seen in Fig. 3.14. Higher current crowding can cause catastrophic LED failures, putting the LEDs' reliability at risk.

The following describes the most possible scenarios that could lead to an EOS/ESD, as well as suggestions for preventing it in the field. In certain cases, customer design has a significant impact on the solution, so it is critical and beneficial for LED manufacturers to collaborate closely with customers to ensure that their fixture designs are EOS/ESD-proof.

### 3.6.1 Electrostatic discharge (ESD)

Operators handling the LED will generate ESD. ESD is a low-energy, short-duration phenomenon that lasts only a few microseconds. It is most commonly generated in non-ESD-protected work environments. For example, the operator's body can have a very different electrical potential than the LED board, resulting in an ESD when they come into contact with LEDs. Readers may want to look at these articles [55,56] for more information on ESD.

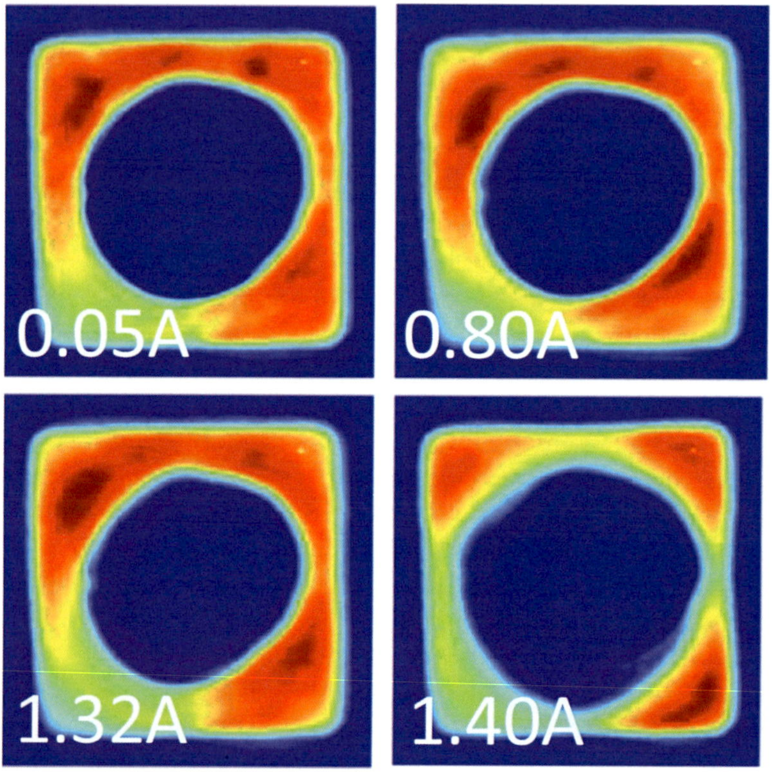


**Figure 3.14** Images of the tested system with a pulse width of 1 ms that are spatially resolved and normalized. The current crowding effect caused by the extremely high current densities reached during the EOS event is visible in the study [38,54]. *(Elsevier. Reprinted with permission.)*

ESD is usually not an issue for LEDs since the majority of LEDs have an ESD suppressor built in to protect the LED chips. However, some of the tiniest new LEDs, which are engineered to optimize light density, lack ESD suppressors. To better secure the LED, these additional components must be placed very close to it. These ESD suppressors have been used to secure other electronic devices for decades and can also be used to protect LEDs.

### 3.6.2 Power supply

EOS could be caused by the luminaire's power supply. Even if the LEDs are not damaged by the user, there are a number of factors that could cause them to fail. Ripple and transient spikes during the turn on and off process, as well as hot plugging, can be silent LED killers, compromising component integrity without causing any noticeable symptoms [31].

It is critical to use power supplies with a low transient peak during the turn on and off phases to avoid EOS failure in the scenarios mentioned above. The maximum rated current of the LED must not be exceeded by these power supplies. Furthermore, the combined typical and ripple currents must not surpass the maximum LED rated current.

Negative pulses or reverse polarity connections of the power supply are another potential cause of failure. This EOS can severely harm LEDs if the operator switches the polarity during the test or development process. To avoid this, use a power supply with short circuit safety and a diode in parallel with the LED string in reverse polarity will be good practices. An assymetrical connector is the best option for correcting mating between the power supply and the connectors on the LED board if the power supply is attached to the LED board with a connector.

## 3.7 Temperature-humidity effect on LEDs and three stages in LED lifetime

Singh et al. used OSRAM high-power LEDs and tested them in a high-temperature-humidity chamber at 85°C and 85% relative humidity to observe the combined temperature-humidity effect on LEDs. The LEDs were placed in the chamber for 24 h before being removed for electrical and optical testing. [57] recorded percentage lumen degradation versus time, and three regions of LED degradation were observed in high temperature and humidity conditions that simulate outdoor applications in an accelerated manner:

**a.** Initial rapid degradation
**b.** Lumen recovery following initial rapid degradation
**c.** Subsequent permanent degradation

As explained in Chapter 1, the initial rapid degradation was caused by light scattering from the LED chip due to moisture penetration into the encapsulant silicone.

There was a recovery after the initial rapid lumen depletion in the first 48 h. As mentioned in Chapter 1, the moisture in silicone was sucked by the die attach, and the reduced moisture content in silicone reduces the scattering of light from the LED chip, resulting in an increase in lumen that acts as though the degradation has been recovered [58].

Following the lumen recovery, the final stage of lumen depletion was observed, with no further recovery. As shown in Fig. 1.7 in Chapter 1 [59], the degradation process started with bond pad contamination and Kirkendall void formation, followed by galvanic dissolution, phosphor dissolution to encapsulant, and die attachment delamination. The breakdown of the Au wire bond causes a shift in resistance, which leads to a reduction in light intensity. In the following chapters, we'll go through each of these stages in depth.

## 3.8 Combined moisture–electrical–thermal effect on LEDs

Tan and Singh suggest a moisture–electrical–temperature (MET) test to assess the outdoor reliability of high-power blue LEDs, with and without phosphor, and to better understand the physics of LED degradation in the presence of the combined humidity, temperature, and electrical stresses [57]. The high-power white LEDs would be blue LEDs with phosphor. When measured under the same environmental conditions, the degradation rate for each LED was found to be different, as shown in Fig. 3.15.

Differences in the coefficient of thermal expansion between the molding component and the lens material, as well as the heat produced by the phosphor layer, are found to account for the major difference in the degradation mechanisms observed for blue and white LEDs, as shown in Table 3.2. When compared to the degradation mechanisms caused by electrical and thermal stress as shown in Table 1.2 in Chapter 1, it is clear that the proposed MET test is needed for the LED industry to evaluate the reliability of LEDs in a real-world outdoor application.

## 3.9 Summary

The impact of high temperatures on LED chips and LED packages was discussed in this chapter. Individual studies on the effect of higher temperature on LED chip and package are addressed, as well as various novel solutions to these problems. The impacts of temperature cycling and temperature shock tests on the overall reliability of LEDs are also addressed, with a case study demonstrating their severity. Following the temperature stress, the effect of electrical overstress, as well as the combined temperature-humidity effect on LEDs, is presented. In the later part of this chapter, the three-stage degradation behavior of LEDs under temperature humidity test is shown, which will be addressed in detail in the following chapters.

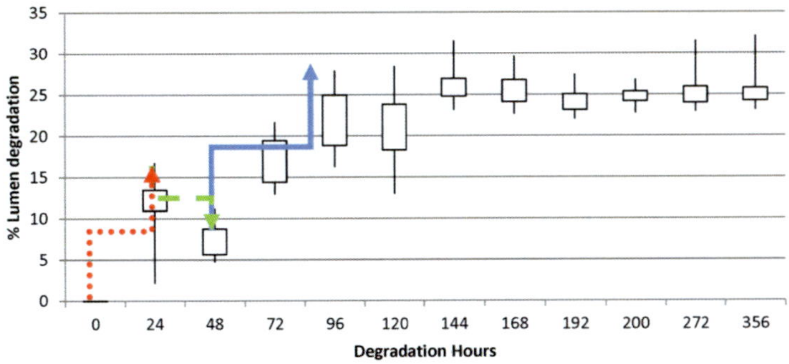


(a)

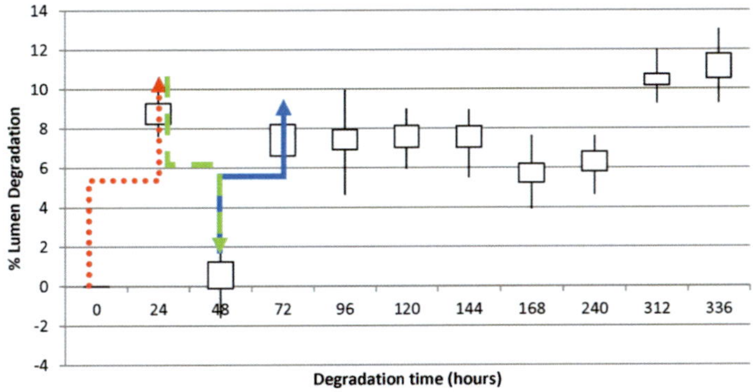


(b)

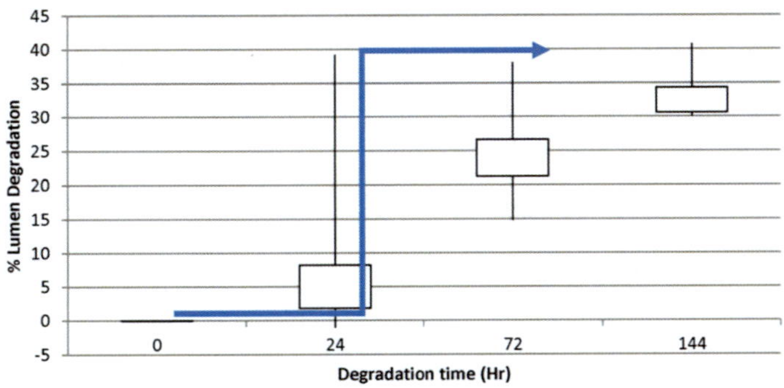


(c)

**Figure 3.15** For blue and white LEDs measured under various operating conditions, percentage lumen degradation versus test time. (a) Blue LEDs in a powered-on state, (b) blue LEDs in a powered-off state, and (c) white LEDs in a powered-on state. For LEDs, the red line (*dotted line*) represents initial degradation, the green line (*dash line*) represents lumen recovery, and the blue line (*solid line*) represents final degradation [38,57]. *(Scientific Reports, Nature. Reprinted with permission.)*

**Table 3.2** Under various operating conditions, the time taken for initial degradation, recovery, and final degradation of blue and white LEDs was measured [57].

| Samples / Stage | Blue LED under on condition | Blue LED under off condition | White LED under on condition | White LED under off condition |
|---|---|---|---|---|
| Initial degradation stage | 12% degradation in first 24 h | 8.5% degradation(in first 24 h | Nil | No degradation |
| Recovery stage | 7.5% degradation till 48 h | 1% degradation till 48 h | Nil | No degradation |
| Final degradation stage | 30% degradation till 356 h | 12% degradation till 336 h | 33% degradation in first 144 h | No degradation |

Scientific Reports, Nature. Reprinted with permission.

## References

[1] M.G. Craford, LEDs for solid state lighting and other emerging applications: status, trends, and challenges, in: Fifth International Conference on Solid State Lighting, vol 5941, International Society for Optics and Photonics, September 2005, p. 594101.
[2] T. Taguchi, Y. Uchida, K. Kobashi, Efficient white LED lighting and its application to medical fields, Physica Status Solidi (A) 201 (12) (2004) 2730–2735.
[3] ITRI, Taiwan. http://iek.iekedu.tw/web/index.php.
[4] http://www.forbes.com/sites/peterdetwiler/2014/03/27/leds-will-get-even-more-efficient-cree-passes-300-lumens-per-watt/#c579d1270b4f.
[5] C. MingTan, P. Singh, A review on degradation physics of high power LEDs in outdoor applications, in: 2016 IEEE 23rd International Symposium on the Physical and Failure Analysis of Integrated Circuits (IPFA), IEEE, July 2016, pp. 383–386.
[6] K. Eichhorn, LEDs in automotive lighting, In Light-Emitting Diodes: Research, Manufacturing, and Applications X 6134 (February 2006) 613405 (International Society for Optics and Photonics).
[7] F.M. Steranka, J. Bhat, D. Collins, L. Cook, M.G. Craford, R. Fletcher, J.J. Wierer, High power LEDs—Technology status and market applications, Physica Status Solidi (A) 194 (2) (2002) 380–388.
[8] N. Narendran, Y. Gu, Life of LED-based white light sources, Journal of Display Technology 1 (1) (2005) 167.
[9] L. Yang, J. Hu, S. Jang, M.W. Shin, Thermal design of ceramic packages for high power light-emitting diodes, Semiconductor Science and Technology 22 (7) (2007) 705.
[10] L.Q. Yang, J.Z. Hu, M.W. Shin, Investigation of thermal measurement variables in high power GaN-based LEDs, Solid State Phenomena 124 (2007) 483–486 (Trans Tech Publications Ltd).
[11] J. Hu, L. Yang, M.W. Shin, Electrical, optical and thermal degradation of high power GaN/InGaN light-emitting diodes, Journal of Physics D: Applied Physics 41 (3) (2008) 035107.

[12] J. Hu, L. Yang, W.J. Hwang, M.W. Shin, Thermal and mechanical analysis of delamination in GaN-based light-emitting diode packages, Journal of Crystal Growth 288 (1) (2006) 157–161.
[13] J. Hu, L. Yang, M.W. Shin, Mechanism and thermal effect of delamination in light-emitting diode packages, Microelectronics Journal 38 (2) (2007) 157–163.
[14] K. Van Doorselaer, K. De Zeeuw, Relation between delamination and temperature-cycling induced failures in plastic packaged devices, in: 40th Conference Proceedings on Electronic Components and Technology, IEEE, May 1990, pp. 813–817.
[15] J. Hu, L. Yang, M.W. Shin, Thermal and mechanical analysis of high-power LEDs with ceramic packages, IEEE Transactions on Device and Materials Reliability 8 (2) (2008) 297–303.
[16] C. DeMilo, C. Bergad, R. Forni, T. Brukilacchio, Thermally induced stresses resulting from coefficient of thermal expansion differentials between an LED sub-mount material and various mounting substrates, In Light-Emitting Diodes: Research, Manufacturing, and Applications XI 6486 (February 2007) 64860N (International Society for Optics and Photonics).
[17] A. Moridi, et al., Residual stresses in thin film systems: effects of lattice mismatch, thermal mismatch and interface dislocations, International Journal of Solids and Structures 50 (22–23) (2013) 3562–3569.
[18] W. Rieger, T. Metzger, H. Angerer, R. Dimitrov, O. Ambacher, M. Stutzmann, Influence of substrate-induced biaxial compressive stress on the optical properties of thin GaN films, Applied Physics Letters 68 (7) (1996) 970–972.
[19] T. Kozawa, T. Kachi, H. Kano, H. Nagase, N. Koide, K. Manabe, Thermal stress in GaN epitaxial layers grown on sapphire substrates, Journal of Applied Physics 77 (9) (1995) 4389–4392.
[20] D.G. Zhao, S.J. Xu, M.H. Xie, S.Y. Tong, H. Yang, Stress and its effect on optical properties of GaN epilayers grown on Si (111), 6H-SiC (0001), and c-plane sapphire, Applied Physics Letters 83 (4) (2003) 677–679.
[21] S.C. Hsu, B.J. Pong, W.H. Li, T.E. Beechem III, S. Graham, C.Y. Liu, Stress relaxation in GaN by transfer bonding on Si substrates, Applied Physics Letters 91 (25) (2007) 251114.
[22] K.S. Stevens, A. Ohtani, A.F. Schwartzman, R. Beresford, Growth of group III nitrides on Si (111) by plasma-assisted molecular beam epitaxy, Journal of Vacuum Science & Technology B: Microelectronics and Nanometer Structures Processing, Measurement, and Phenomena 12 (2) (1994) 1186–1189.
[23] S.C. Hsu, C.Y. Liu, Stress analysis of transferred thin-GaN LED by Au-Si wafer bonding, in: Fifth International Conference on Solid State Lighting, vol 5941, International Society for Optics and Photonics, September 2005, p. 594116.
[24] T.L. Chou, C.F. Huang, C.N. Han, S.Y. Yang, K.N. Chiang, Fabrication process simulation and reliability improvement of high-brightness LEDs, Microelectronics Reliability 49 (9–11) (2009) 1244–1249.
[25] A. Cremades, L. Görgens, O. Ambacher, M. Stutzmann, F. Scholz, Structural and optical properties of Si-doped GaN, Physical Review B 61 (4) (2000) 2812.
[26] L. Tan, J. Li, K. Wang, S. Liu, Effects of defects on the thermal and optical performance of high-brightness light-emitting diodes, IEEE Transactions on Electronics Packaging Manufacturing 32 (4) (2009) 233–240.
[27] A.T. DiBenedetto, Prediction of the glass transition temperature of polymers: a model based on the principle of corresponding states, Journal of Polymer Science B: Polymer Physics 25 (9) (1987) 1949–1969.
[28] K. Zhang, M.M.F. Yuen, Thermal stress in LED packages for solid-state lighting, in: R.B. Hetnarski (Ed.), Encyclopedia of Thermal Stresses, Springer, Dordrecht, 2014, https://doi.org/10.1007/978-94-007-2739-7_272.

[29] J.J. Licari, D.W. Swanson, Adhesives Technology for Electronic Applications: Materials, Processing, Reliability, William Andrew, 2011.
[30] J. Lau (Ed.), Thermal Stress and Strain in Microelectronics Packaging, Springer Science & Business Media, 2012.
[31] C. Tsou, Y.S. Huang, Silicon-based packaging platform for light-emitting diode, IEEE Transactions on Advanced Packaging 29 (3) (2006) 607–614.
[32] C. Weng, W.B. Lee, S. To, Birefringence techniques for the characterization of residual stresses in injection-moulded micro-lens arrays, Polymer Testing 28 (7) (2009) 709–714.
[33] M.H. Chang, D. Das, P.V. Varde, M. Pecht, Light emitting diodes reliability review, Microelectronics Reliability 52 (5) (2012) 762–782.
[34] F. Wu, W. Zhao, S. Yang, C. Zhang, Failure modes and failure analysis of white LEDs, in: 2009 9th International Conference on Electronic Measurement & Instruments, IEEE, August 2009, pp. 4–978.
[35] J. Arnold, When the Lights Go Out: LED Failure Modes and Mechanisms, white paper, Dfr Solutions, college park, Margland, 2004.
[36] G. Lu, S. Yang, Y. Huang, Analysis on failure modes and mechanisms of LED, in: 2009 8th International Conference on Reliability, Maintainability and Safety, IEEE, July 2009, pp. 1237–1241.
[37] N.Y. Shammas, Present problems of power module packaging technology, Microelectronics Reliability 43 (4) (2003) 519–527.
[38] Z. Chen, Q. Zhang, R. Chen, F. Jiao, M. Chen, X. Luo, S. Liu, Comparison of LED package reliability under thermal cycling and thermal shock conditions by experimental testing and finite element simulation, in: 2011 IEEE 61st Electronic Components and Technology Conference (ECTC), IEEE, May 2011, pp. 454–459.
[39] M. Meneghini, L.R. Trevisanello, G. Meneghesso, E. Zanoni, A review on the reliability of GaN-based LEDs, IEEE Transactions on Device and Materials Reliability 8 (2) (2008) 323–331.
[40] M. Meneghini, A. Tazzoli, G. Mura, G. Meneghesso, E. Zanoni, A review on the physical mechanisms that limit the reliability of GaN-based LEDs, IEEE Transactions on Electron Devices 57 (1) (2009) 108–118.
[41] M. Meneghini, L.R. Trevisanello, U. Zehnder, T. Zahner, U. Strauss, G. Meneghesso, E. Zanoni, High-temperature degradation of GaN LEDs related to passivation, IEEE Transactions on Electron Devices 53 (12) (2006) 2981–2987.
[42] X. Shao, H. Lu, D. Chen, Z. Xie, R. Zhang, Y. Zheng, Efficiency droop behavior of direct current aged GaN-based blue light-emitting diodes, Applied Physics Letters 95 (16) (2009) 163504.
[43] S. Buso, G. Spiazzi, M. Meneghini, G. Meneghesso, Performance degradation of high-brightness light emitting diodes under DC and pulsed bias, IEEE Transactions on Device and Materials Reliability 8 (2) (2008) 312–322.
[44] F. Manyakhin, A. Kovalev, A.E. Yunovich, Aging mechanisms of InGaN/AlGaN/GaN light-emitting diodes operating at high currents, Materials Research Society Internet Journal of Nitride Semiconductor Research 3 (1998).
[45] F. Rossi, M. Pavesi, M. Meneghini, G. Salviati, M. Manfredi, G. Meneghesso, E. Zanoni, Influence of short-term low current dc aging on the electrical and optical properties of InGaN blue light-emitting diodes, Journal of Applied Physics 99 (5) (2006) 053104.
[46] L. Liu, M. Ling, J. Yang, W. Xiong, W. Jia, G. Wang, Efficiency degradation behaviors of current/thermal co-stressed GaN-based blue light emitting diodes with vertical-structure, Journal of Applied Physics 111 (9) (2012) 093110.

[47] G. Meneghesso, S. Levada, E. Zanoni, S. Podda, G. Mura, M. Vanzi, I. Eliashevich, Failure modes and mechanisms of DC-aged GaN LEDs, Physica Status Solidi (A) 194 (2) (2002) 389–392.
[48] L. Huang, T. Yu, Z. Chen, Z. Qin, Z. Yang, G. Zhang, Different degradation behaviors of InGaN/GaN MQWs blue and violet LEDs, Journal of Luminescence 129 (12) (2009) 1981–1984.
[49] S. Bychikhin, D. Pogany, L.K.J. Vandamme, G. Meneghesso, E. Zanoni, Low-frequency noise sources in as-prepared and aged GaN-based light-emitting diodes, Journal of Applied Physics 97 (12) (2005) 123714.
[50] S. Tomiya, T. Hino, S. Goto, M. Takeya, M. Ikeda, Dislocation related issues in the degradation of GaN-based laser diodes, IEEE Journal of Selected Topics in Quantum Electronics 10 (6) (2004) 1277–1286.
[51] S.N. Lee, H.S. Paek, J.K. Son, H. Kim, K.K. Kim, K.H. Ha, Y. Park, Effects of Mg dopant on the degradation of InGaN multiple quantum wells in AlInGaN-based light emitting devices, Journal of Electroceramics 23 (2–4) (2009) 406.
[52] M. Meneghini, S. Podda, A. Morelli, R. Pintus, L. Trevisanello, G. Meneghesso, E. Zanoni, High brightness GaN LEDs degradation during dc and pulsed stress, Microelectronics Reliability 46 (9–11) (2006) 1720–1724.
[53] M. Meneghini, L. Trevisanello, S. Podda, S. Buso, G. Spiazzi, G. Meneghesso, E. Zanoni, Stability and performance evaluation of high-brightness light-emitting diodes under DC and pulsed bias conditions, in: Sixth International Conference on Solid State Lighting, vol 6337, International Society for Optics and Photonics, September 2006, p. 63370R.
[54] https://www.led-professional.com/resources-1/articles/electrical-over-stress-how-to-prevent-an-led-failing-earlier-than-expected.
[55] M.D. Ker, K.C. Hsu, Overview of on-chip electrostatic discharge protection design with SCR-based devices in CMOS integrated circuits, IEEE Transactions on Device and Materials Reliability 5 (2) (2005) 235–249.
[56] M.D. Ker, H.H. Chang, How to safely apply the LVTSCR for CMOS whole-chip ESD protection without being accidently triggered on, Journal of Electrostatics 47 (4) (1999) 215–248.
[57] P. Singh, C.M. Tan, Degradation physics of high power LEDs in outdoor environment and the role of phosphor in the degradation process, Scientific Reports 6 (1) (2016) 1–13.
[58] P. Singh, C.M. Tan, L.B. Chang, Early degradation of high power packaged LEDs under humid conditions and its recovery—myth of reliability rejuvenation, Microelectronics Reliability 61 (2016) 145–153.
[59] C.M. Tan, P. Singh, Time evolution degradation physics in high power white LEDs under high temperature-humidity conditions, IEEE Transactions on Device and Materials Reliability 14 (2) (2014) 742–750.

## Further reading

[1] L. Tan, J. Li, Z. Liu, K. Wang, P. Wang, Z. Gan, S. Liu, A light emitting diode's chip structure with low stress and high light extraction efficiency, in: 2008 58th Electronic Components and Technology Conference, IEEE, May 2008, pp. 783–788.
[2] L. Trevisanello, M. Meneghini, G. Mura, M. Vanzi, M. Pavesi, G. Meneghesso, E. Zanoni, Accelerated life test of high brightness light emitting diodes, IEEE Transactions on Device and Materials Reliability 8 (2) (2008) 304–311.

[3] M.D. Ker, H.H. Chang, Cascoded LVTSCR with tunable holding voltage for ESD protection in bulk CMOS technology without latchup danger, Solid-State Electronics 44 (3) (2000) 425–445.
[4] C.M. Tan, B.K. Chen, X. Li, S.J. Chen, Rapid light output degradation of GaN-based packaged LED in the early stage of humidity test, IEEE Transactions on Device and Materials Reliability 12 (1) (2011) 44–48.

# CHAPTER 4

# Initial rapid lumen degradation for high-power white packaged LEDs

## 4.1 Introduction

The effect of humidity on LED characteristics is discussed in Chapter 3, and the unique degradation pattern of LEDs where rapid lumen degradation in the first few hours of their lifetime followed by a recovery phase is mentioned. Such degradation pattern does not happen to integrate circuits when exposed to humidity [1]. This chapter delves into the reasons for such rapid lumen degradation in high-power white LEDs in the first few hours.

Tan et al. [2] tested 15 OSRAM 3-W Platinum DRAGON white packed LEDs in humidity chamber at 85°C/85% RH relative humidity, per IPC/JEDEC standard. All the LEDs were turned off during testing to prevent the evaporation of the trapped moisture from the encapsulation due to the heating from the LED chip during testing.

The LEDs were removed from the humidity chamber at specific times for electrical measurement and optical measurement. The specific test times were set at 0, 48, 96, 192, 384, 1000, and 2500 h for practical consideration. A 700-mA pulsed train (500 ms pulse width with 50% duty cycle) was applied to the LEDs during the optical measurement to prevent self-heating during measurement, according to Ref. [3].

### 4.1.1 Experimental test results

Time variation of LEDs' luminous flux degradation was summarized in Fig. 4.1, and the lumen decreases with time as a result of the prolonged humidity test, as predicted. However, within the first 48 h of the accelerated humidity test, all of the LEDs possessed rapid lumen drop ranging from 0.71% to 2.22%, with an average drop of 1.32%. Compared to the mean drop of 1.59% and 1.93% after 96 and 192 h of accelerated testing

*Reliability and Failure Analysis of High-Power LED Packaging*
ISBN 978-0-12-822408-3
https://doi.org/10.1016/B978-0-12-822408-3.00005-8

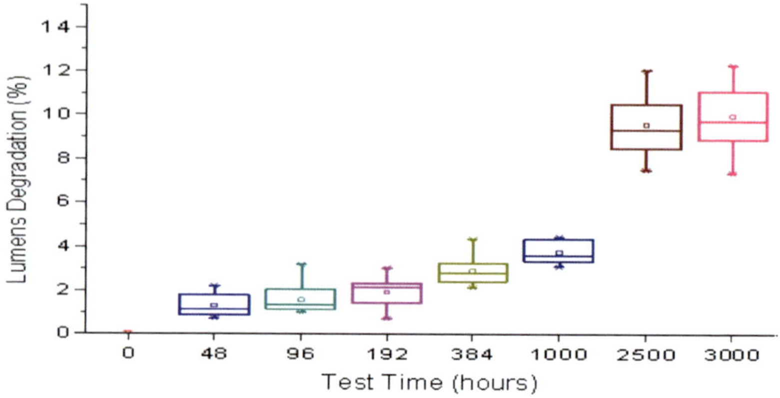


**Figure 4.1** Lumen degradation (in %) of white GaN-based LEDs subjected to 85°C/85% RH [2]. *(Reprinted with permission, IEEE Transactions on Computer and Materials Efficiency, IEEE Transactions on.)*

respectively, this initial rapid decrease in lumen is a substantial contribution to the LED's lumen degradation over its lifespan under humid conditions. Tan et al. [2] explored the underlying degradation processes that cause this initial rapid decline.

## 4.2 Physical analysis on the degraded samples

For LED degradation, the possible sources will be die-related, phosphor-related, and the silicone encapsulation. In comparing the forward I—V characteristics of the LEDs obtained at 0 h and after 48 h of testing, no significant change in their I—V curve was observed [2] as shown in Fig. 4.2. As the ideality factor (n) and reverse saturation current (Is) can be used to describe the characteristics of any p-n-junction, they were computed from their I—V curve, and the results are displayed in Tables 4.1 and 4.2. The calculated ideality factors range from 3.1 to 4, within the recommended range of 2.0—7.0 [4].

The ideality factors of the LEDs at 0 and 48 h have a difference of less than 1% as shown in Table 4.1. On the other hand, very large difference was observed in reverse saturation currents at 0 and 48 h, and they ascribed such difference to the computation and noise embedded in the measurement, due to the femtoamperes order of the current. Overall, die-related failure was ruled out as a cause of the rapid degradation.

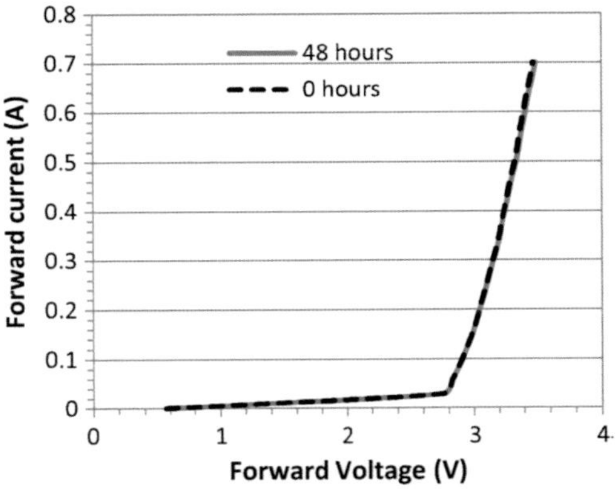


**Figure 4.2** Forward I–V characteristics of the LEDs at 0 and 48 h of 85% RH/85C testing [2]. *(Reprinted with permission, IEEE Transactions on Computer and Materials Efficiency, IEEE Transactions on.)*

**Table 4.1** Ideality variables of five LEDs in the 85/85 test at 0 and 48 h [2].

| Unit # | 1 | 2 | 3 | 4 | 5 |
|---|---|---|---|---|---|
| 0 h | 3.64 | 3.86 | 3.24 | 3.13 | 3.77 |
| 48 h | 3.61 | 3.84 | 3.23 | 3.12 | 3.78 |
| % Change | 0.82 | 0.52 | 0.31 | 0.32 | −0.10 |

Reprinted with permission, IEEE Transactions on Computer and Materials Efficiency, IEEE Transactions on.

**Table 4.2** Reverse saturation currents (in fA) of five LEDs at 0 and 48 h of 85°C/85% testing [2].

| Unit # | 1 | 2 | 3 | 4 | 5 |
|---|---|---|---|---|---|
| 0 h | 8.38 | 46.74 | 0.27 | 0.05 | 26.2 |
| 48 h | 6.72 | 42.33 | 0.23 | 0.05 | 27.4 |
| % Change | 19.8 | 9.43 | 14.7 | 7 | −4.6 |

Reprinted with permission, IEEE Transactions on Computer and Materials Efficiency, IEEE Transactions on.

The blue-to-yellow ratio of the light from the LEDs at 0 and 48 h also indicated no noticeable variations, suggesting that phosphor failure was also impossible. In fact, no change in CRI of the LEDs between 0 and 48 h was observed, and it was only 0.06% change after 96 h of exposure under the 85/85 test condition.

From their analysis, and the fact that the light output from the LED chip must pass through the transparent encapsulation, the rapid light output degradation observed should therefore be linked to the encapsulation. Silicone was used to encapsulate the LED chip. Silicone is commonly used for LEDs today providing protection to the interconnecting wires and the chip while improving their optical performance. However, since silicone encapsulation cannot provide a hermetic seal, moisture can diffuse through it during humidity test [5].

## 4.3 Thermal gravimetric analysis (TGA) analysis

To investigate moisture diffusion and entrapment in silicone encapsulation, Tan et al. used TGA technique on eight LEDs. Samples #1 and #2 were unstressed LEDs, #3—#5 were subjected to 24 h of 85/85 test, and #6—#8 were subjected to 48 h of 85/85 test. The temperature of each LED was raised in the TGA instrument from 30 to 170°C at a rate of 3°C per minute in an ultrahigh-purity nitrogen environment with nitrogen purge rate into the instrument at 40 mL/min, the weight percentage changes of the LED with temperature were used for the determination of the amount of moisture ingression into the silicone encapsulation during the humidity test. TGA temperature was set to 170°C so that the weight loss of the LED was solely due to the entrapped moisture [6].

Fig. 4.3 indicates that all the stressed LEDs indeed have substantially higher moisture desorption than unstressed LEDs, and the water desorption amount for the LEDs exposed for 24 and 48 h was the same. It is clear that moisture was entrapped in the LEDs after stressed, and the moisture entrapment levels were saturated after 24 h. Furthermore, majority of the moisture desorption occurred at temperatures above 90°C, implying that the entrapped moisture was unlikely to be driven out during normal LED activity.

## 4.4 Energy dispersive system (EDS) analysis

To examine moisture entrapment profile in silicone under humidity exposure qualitatively, EDS was applied on LEDs that were subjected to 24-h humidity test. To probe the presence of moisture in silicone encapsulation at different depths, different accelerating voltages of 10, 15, and 20 keV were employed.

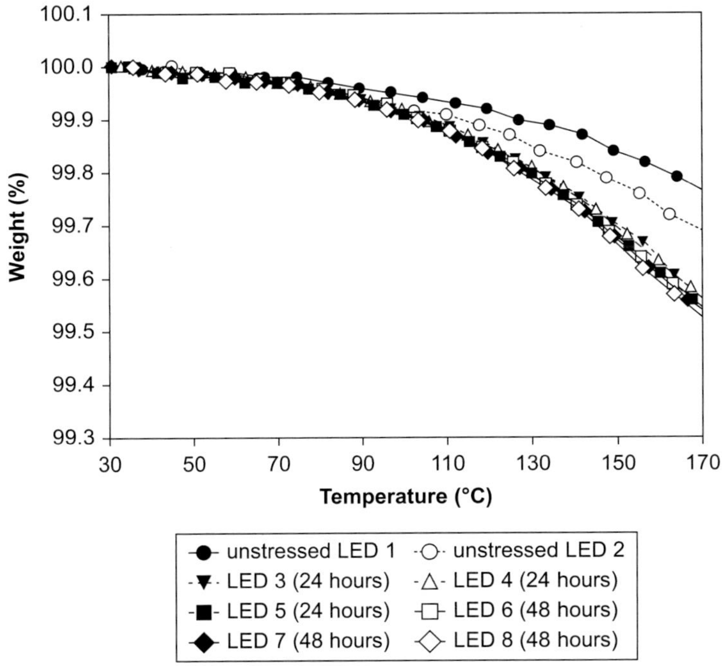


**Figure 4.3** Shows how the LED's weight shifts as a function of temperature (in percent) [2]. *(Reprinted with permission, IEEE Transactions on Computer and Materials Efficiency, IEEE Transactions on.)*

Hydrogen cannot be detected by EDS, but moisture contains oxygen, the presence of moisture can thus be detected qualitatively by comparing the oxygen content to the baseline silicon content in silicone. Fig. 4.4 shows the normalized ratio of the atomic percentage of oxygen over that of silicon for stressed LEDs, using the unstressed LEDs acting as the normalization baseline. As compared to the unstressed silicone, a higher ratio of oxygen content to silicon content in the silicone of the stressed LED implies a higher amount of oxygen factor and thus the presence of moisture. Tan et al. [2] obtained the depth profile of moisture in the silicone encapsulation qualitatively by increasing the accelerating voltage of the SEM.

At 25 KeV, the EDS spectrum did not show any element from phosphor, hence it can be concluded that the moisture did not reach the die surface. This is consistent with the finding that both the LED chip and phosphor layer were not affected in this rapid degradation period.

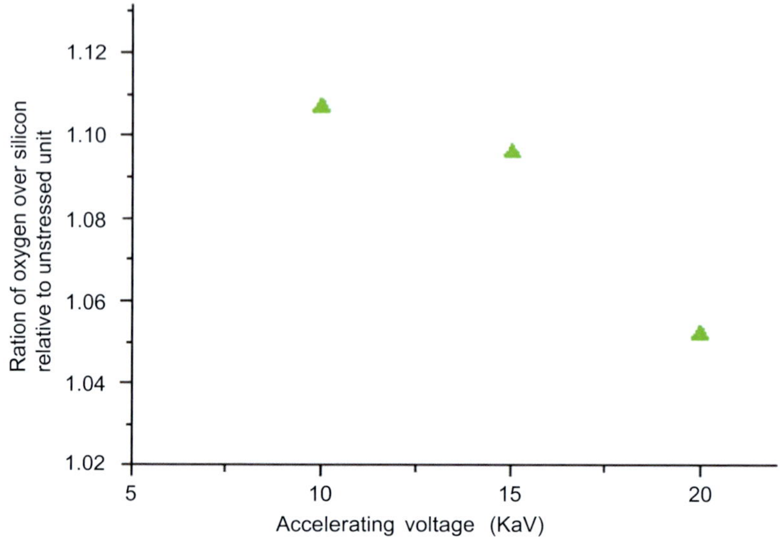


**Figure 4.4** Ratio of oxygen (atomic percentage) over silicon (atomic percentage) from EDS analysis at different accelerating voltage for an LED stressed for 48 h [2]. *(Reprinted with permission, IEEE Transactions on Computer and Materials Efficiency, IEEE Transactions on.)*

## 4.5 Modeling analysis

Based on the thorough investigation mentioned above, it can be concluded that the rapid light output degradation of the white LEDs in the early stages of the humidity test was most likely due to the entrapment of moisture in the silicone material. Tan et al. [2] believed that this entrapped moisture could scatter or absorb some light from the LED die and the phosphor, contributing to the sharp drop in the light output. They verified their results using light scattering simulation.

As shown in Fig. 4.5, they modeled the entrapped moisture in the silicone encapsulation as embedded particles that would scatter or absorb the light coming from the LED chip and phosphor. The material, size, and shape of the particles affected the angular distribution of scattered light. Their SEM images indicated that the pore size in silicone was smaller than the scale of the visible light wavelength, and therefore the embedded water particles were likewise. Hence they applied Rayleigh's scattering [7] for their analysis.

For simplicity and qualitative verification, Tan et al. [8] assumed that the moistures were evenly distributed in the silicone. This slab of water particles

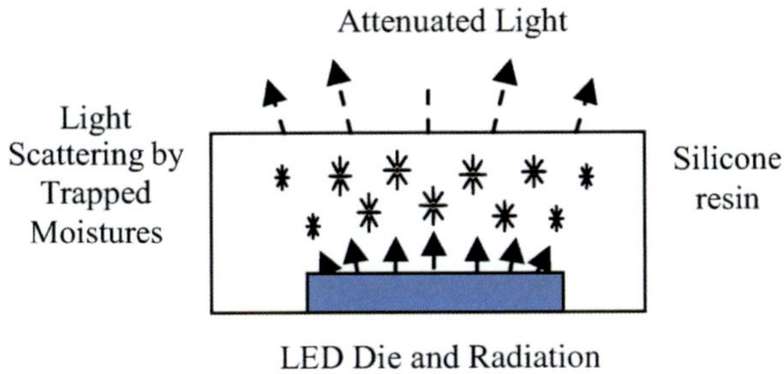


**Figure 4.5** Light scattering in silicone resin caused by trapped moisture [2]. *(Reprinted with permission, IEEE Transactions on Computer and Materials Efficiency, IEEE Transactions on.)*

will scatter or absorb light, reducing its intensity. The light scattering cross section is given as follows.

$$\sigma_{scat} = \frac{8\pi}{3}\left(\frac{2\pi\, n_{sl}}{\lambda}\right)^4 r^6 \left(\frac{m^2-1}{m^2+2}\right)^2 \tag{4.1}$$

and the absorption cross section is given approximately as [7].

$$\sigma_{abs} = \frac{4\pi k}{\lambda}\,\frac{9m}{(m^2+2)^2} \tag{4.2}$$

The attenuated irradiance by the slab of water particles is equal to

$$I_t = I_0 \exp(-\rho\,\sigma_{extinction} t) = I_0 \exp[-\rho(\sigma_{scat}+\sigma_{abs})t] \tag{4.3}$$

where $r$ is the particle's radius, $I_0$ is the irradiance of nonpolarized incident light, i.e., the light output from the die/phosphor, $\rho$ is the number of particles per unit volume, $t$ is the depth of the water slab entrapped in the silicone, and $m$ is the relative complex refractive index computed as:

$$m = \frac{n_{wt}+ik_{wt}}{(n_{sl}+ik_{sl})} = \frac{n_{wt}+i\left(\alpha_{wt}\dfrac{\lambda}{4\pi}\right)}{n_{sl}+i\left(\alpha_{sl}\dfrac{\lambda}{4\pi}\right)} \tag{4.4}$$

Here, $n_{wt}$ and $n_{sl}$ are the refractive indexes, $k_{wt}$ and $k_{sl}$ are the extinction coefficients, and $\alpha_{wt}$ and $\alpha_{sl}$ are the absorption coefficients of water and silicone resin, respectively. $\rho$ is related to the volume fraction of water in silicone given as

$$\rho = \frac{\text{volume of water}}{(\text{volume of single water particle in the resin})} = \frac{\text{volume fraction} \times 1\ \mu m^3}{(1.33\pi r^3)} \tag{4.5}$$

It was discovered that the volume fraction of water in silicone saturated quickly at around 0.064% [9], which clarified the saturation of moisture absorption seen in TGA results.

## 4.6 Degradation mechanism

Since the white light from the LEDs in their work was a mixture of blue light from the die and yellow light from the phosphor on the die, their radiation spectrum has two local maxima. The blue light had a narrow distribution with peak at 460 nm, while the wavelength of the yellow light had a broad distribution of 550–590 nm. According to OSRAM's data sheet, the luminous flux ratios of photometric intensity between the two colors were 9% (blue) and 91% (yellow). At blue and yellow light wavelengths, water had refractive indexes of 1.339 and 1.334, respectively [10], while silicone had refractive indexes of 1.532 and 1.519, respectively [11]. A Nikon measuring microscope showed that the silicone thickness was 0.52 mm (MM-40).

Water absorption coefficients for blue and yellow lights are 0.0002/cm and 0.001/cm, respectively [12], while silicone absorption coefficients ranged from 70/cm to 700/cm [13]. Hence negligible absorption of the emitted light by water particles in the silicone is expected, and light scattering should be the main contributor to the attenuation.

The amount of the entrapped moisture determines the volume of light extinction. For moisture particle diameter of 50 nm, the volume fraction is 0.064%, and the extinction of yellow light is 0.56%. The attenuation increases to 4.4% when the particle size becomes 100 nm. Fig. 4.6 shows the percentage attenuation of blue and yellow lights as a function of the particle size (i.e., silicone pore size). As the size of the pores varies and the moisture distribution in the silicone is not uniform, as shown in Fig. 4.4, attenuated irradiance will change according to Eq. (4.3), resulting in the observed difference in light attenuation during the initial period of rapid lumen degradation.

Fig. 4.6 shows that the percentage attenuation increases dramatically when the pore size exceeds 50 nm. Negligible attenuation is expected

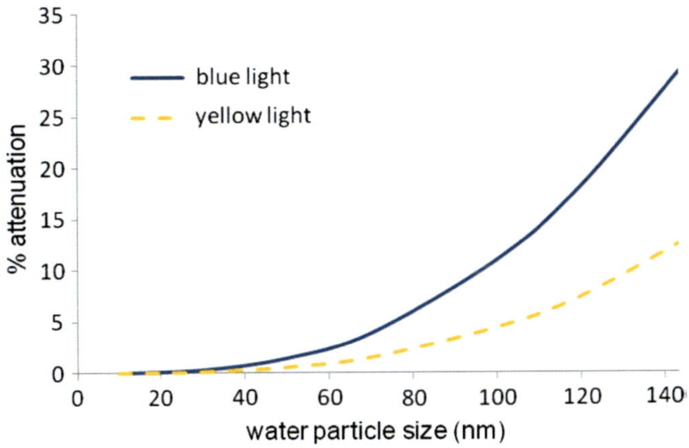


**Figure 4.6** Attenuation of blue and yellow lights with respect to water particle size [2]. *(Reprinted with permission, IEEE Transactions on Computer and Materials Efficiency, IEEE Transactions on.)*

when the pore size is less than 30 nm. Consequently, sharp drop in light production from LED in a humid environment can be prevented if the pore size of silicone is less than 50 nm.

Shuto et al. [9] discovered that moisture content becomes saturated in 2 mm thick silicone after approximately 2 h of humidity exposure at 85% relative humidity and 85°C. As the silicone was just 0.52 mm thick in the LEDs studied by Tan et al. [2], the time to moisture saturation in the silicone will be just 8 min at 85°C/85%RH. Thus, the finding of the same amount of moisture in the silicone when the LEDs were exposed to humidity for 24 and 48 h by Tan et al. [2] is consistent.

### 4.6.1 Dry temperature on LEDs degradation

Tan et al. [2] tested the aforementioned hypothesis by subjecting three LEDs from the same batch to a constant temperature of 85°C with humidity at ambient RH, which is in the range of 30%–40%. The LED luminous flux degradation (in percent) up to 48 h is shown in Table 4.3.

From Table 4.3, the degradation rates are significantly different among the LEDs. The average decrease in luminous flux was just 0.355% after 48 h of high temperature exposure. This is much lower than the average sharp decrease of 1.32% in the case of 85°C/85% RH. The deterioration of LEDs #1 and #2 proceeded at a slow rate, which cannot be called a sudden decrease. For LED #3, the deterioration stops after 24 h with a drop of 0.268%. In other words, these findings back up the theory proposed by Tan

**Table 4.3** Lumen flux degradation of three LEDs from the same batch subjected to a constant temperature of 85°C with ambient humidity [2].

| Exposure Duration | % Luminous flux degradation | | |
|---|---|---|---|
| | LED #1 | LED #2 | LED #3 |
| 24 h | 0.35 | 0.16 | 0.26 |
| 48 h | 0.6 | 0.19 | 0.26 |



et al. [2] that the sudden drop was caused by moisture entrapment. Such finding is also found to be used in the health monitoring of LEDs using machine learning [14].

## 4.7 Summary

The initial sharp drop in LED light production was found to be the moisture entrapment in the silicone encapsulation. Light scatter from these entrapped moistures as the white light passed through the encapsulation. It is also shown that once moisture has penetrated deep into the silicone, normal LED operation can unlikely force the moisture out of the encapsulation, resulting in a permanent deterioration of the LED performance and a reduction in its efficacy. The pore size of the silicone should be less than 50 nm to avoid such initial rapid light degradation.

## References

[1] C.M. Tan, P. Singh, Time evolution degradation physics in high power white LEDs under high temperature-humidity conditions, IEEE Transactions on Device and Materials Reliability 14 (2) (2014) 742–750.

[2] C.M. Tan, B.K. Chen, X. Li, S.J. Chen, Rapid light output degradation of GaN-based packaged LED in the early stage of humidity test, IEEE Transactions on Device and Materials Reliability 12 (1) (2011) 44–48.

[3] C.M. Tan, Z. Gan, W.F. Ho, S. Chen, R. Liu, Determination of the dice forward *I*– *V* characteristics of a power diode from a packaged device and its applications, Microelectrontronics Reliability 45 (1) (January, 2005) 179–184.

[4] J.M. Shah, Y.-L. Li, T. Gessmann, E.F. Schubert, Experimental analysis and theoretical model for anomalously high ideality factors ($n$ _ 2.0) in AlGaN/GaN p-n junction diodes, Journal of Applied Physics 94 (4) (August, 2003) 2627–2630.

[5] J.S. Tonge, M. Mohamed, N.E. Shepherd, B.M. Ketola, Reliability of silicone encapsulants in high pollution environments, in: Conf. Rec. IEEE 4th World Conf. Photovolt. Energy Convers, 2006, pp. 2193–2194.

[6] Y. Shi, A.A.O. Tay, E.H. Wong, R. Ranjan, An effective method of characterizing moisture desorption of polymeric materials at high temperature, in: Proc. 4th Electron. Packag. Technol. Conf, 2002, pp. 70–75.

[7] C.F. Bohren, D.R. Huffman, Absorption and Scattering of Light by Small Particles, Wiley, Hoboken, NJ, 1998.
[8] A.J. Cox, A.J. Deweerd, J. Linden, An experiment to measure Mie and Rayleigh total scattering cross sections, American Journal of Physics 70 (6) (June, 2002) 620−625.
[9] Y. Shuto, H. Sato, M. Fukuda, S. Tohno, Optical, thermal-strain, and water-sorption properties of plastic resins for laser-diode molding, Electronics and Communications in Japan 81 (12) (December, 1998) 17−26.
[10] G.M. Hale, M.R. Querry, Optical constants of water in the 200-nm to 200-microm wavelength region, Applied Optics 12 (3) (March 1, 1973) 555−563.
[11] B. Riegler, R. Thomaier, Index-matching Silicones Enable High Brightness LED Packages, February, 2006. LEDs Mag.
[12] R.C. Smith, K.S. Baker, Optical properties of the clearest natural waters (200−800 nm), Applied Optics 20 (2) (January, 1981) 177−184.
[13] P.J. Mendoza, Combined photothermal and photoacoustic characterization of silicon-epoxy, and the existence of a particle percolation threshold, Analytical Sciences 17 (April, 2001) 269−272.
[14] M.S. Ibrahim, Z. Jing, J. Fan, Health monitoring, machine learning, and digital twin for LED degradation analysis, in: Reliability of Organic Compounds in Microelectronics and Optoelectronics, Springer, Cham, 2022, pp. 151−210.

## Further reading

[1] R. Baillot, Y. Deshayes, L. Bechou, T. Buffeteau, I. Pianet, C. Armand, F. Voillot, S. Sorieul, Y. Ousten, Effects of silicone coating degradation on GaN MQW LEDs performances using physical and chemical analyses, Microelectronics Reliability 50 (9−11) (September−November, 2010) 1568−1573.
[2] Z. Qin, X. Mu, K. Wang, Z. Gan, X. Luo, S. Liu, Dynamic mechanical properties of the transparent silicone resin for high power LED packaging, in: Proc. ICEPT-HDP, 2008, pp. 1−4.

CHAPTER 5

# Lumen recovery in high-power LEDs under prolonged outdoor operation

## 5.1 Introduction

Many studies have been done on the impact of high temperature or electrical effects on LED degradation trends, but only a few studies have been done on the impact of moisture on LED degradation, as stated in previous Chapters. LEDs are currently evaluated using international standards developed to assess IC humidity reliability [1,2], but Tan and Singh indicated that such standards are not sufficient for studying LED humidity reliability [3].

Tan et al. observed a unique degradation trend in high-power LEDs under humid conditions, as discussed in previous chapters. A rapid lumen degradation is observed at first, followed by a recovery. The initial rapid lumen degradation mechanism was well defined in Chapter 4, and the aim of this chapter is to look into the physics of lumen recovery.

Singh et al. used 40 units of high-power OSRAM golden dragon white LEDs that were mounted in a high temperature-humidity chamber to conduct accelerated life research. The test temperature was set to 85°C, and two moisture levels, 85% RH and 95% RH, were selected, each with a sample size of 20 LEDs (with no bias). [4] offers supplementary detail on test configuration and measurement techniques.

After every 24 h, the LEDs' luminous flux was measured, and the percentage luminous flux difference over time was plotted as a box-plot for the two sets of LEDs, as shown in Figs. 5.1 and 5.2. Both sets of LEDs tested at different moisture levels showed rapid lumen degradation, as predicted. This initial rapid lumen degradation reaches a peak value, followed by a recovery in lumen degradation. The initial rapid lumen degradation reaches its peak at different time intervals, i.e., 120 and 72 h, respectively, for

*Reliability and Failure Analysis of High-Power LED Packaging*
ISBN 978-0-12-822408-3
https://doi.org/10.1016/B978-0-12-822408-3.00003-4

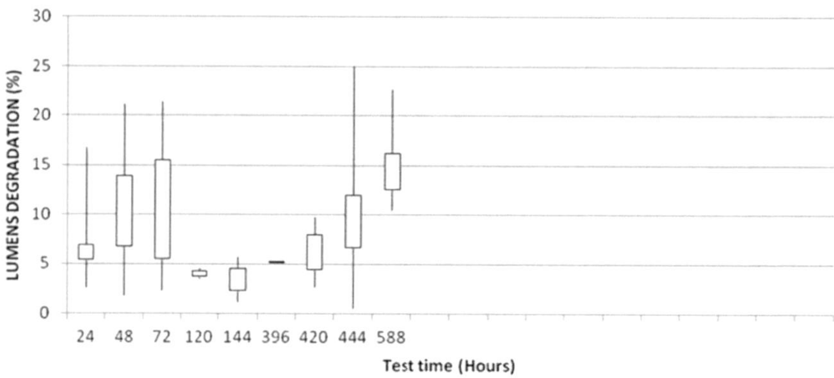


**Figure 5.1** Indicates the percentage luminous flux deterioration over time for the LEDs with a 95% RH [5]. *(Reprinted with permission, © Microelectronics Reliability, Elsevier.)*

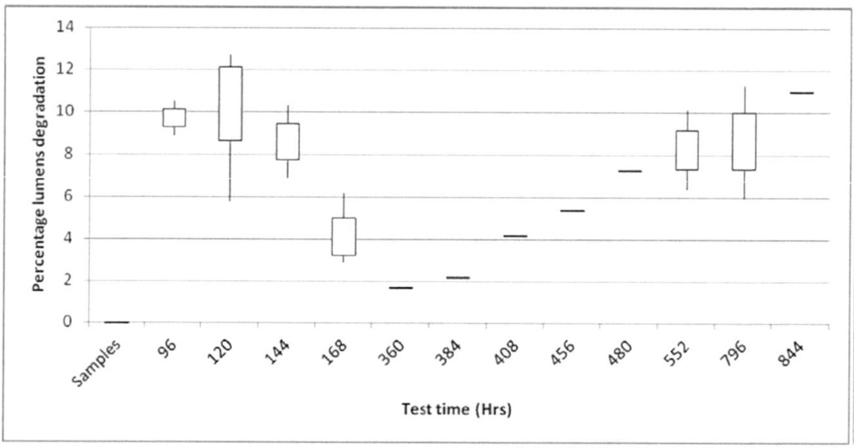


**Figure 5.2** Indicates the percentage luminous flux deterioration over time for the LEDs with an RH of 85% [5]. *(Reprinted with permission, © Microelectronics Reliability, Elsevier.)*

samples measured under 85% RH and 95% RH. At higher moisture levels, the peak value is higher.

## 5.2 Statistical analysis on the test results

The Akaike information criterion (AIC) was used by Singh et al. [5] to calculate the number of possible failure mechanisms during the first two stages of LED degradation, i.e., rapid degradation followed by "recovery period." To evaluate the probability of each unit belonging to each failure process, they used simulated annealing (SA) in combination with expectation—maximization (EM). Refs. [4,6—9] provide more information on the various electronic systems and its effective implementations. They discovered that only one failure mechanism was present in each stage of the LED degradation trend using the SA and EM methods, as shown in Tables 5.1 and 5.2.

They calculated the parameters values of the statistical distributions associated with the degradation mechanisms during the two phases, as shown in Table 5.3, after deciding that there is only one degradation mechanism. Table 5.3 clearly shows that shape parameters have different values, indicating that different mechanisms are responsible for LED failure at different levels, as predicted by Singh et al. [5].

**Table 5.1** Under 95% RH/85°C, the time to achieve maximum initial degradation before recovery stage for each LED, as well as the likelihood of each unit belonging to the degradation mechanism. The units are numbered in order of increasing time to initial degradation [5].

| Unit # | TTF (time to failure) | Inlier (I)/Outlier (O) | Pr (FM 1) |
|---|---|---|---|
| 4 | 24 | I | 1 |
| 5 | 24 | I | 1 |
| 8 | 24 | I | 1 |
| 10 | 48 | I | 1 |
| 12 | 48 | I | 1 |
| 13 | 48 | I | 1 |
| 15 | 48 | I | 1 |
| 17 | 48 | I | 1 |
| 1 | 72 | I | 1 |
| 2 | 72 | I | 1 |
| 6 | 72 | I | 1 |
| 9 | 72 | I | 1 |
| 11 | 72 | I | 1 |
| 14 | 72 | I | 1 |
| 19 | 72 | I | 1 |
| 20 | 72 | I | 1 |
| 16 | 96 | I | 1 |
| 18 | 96 | I | 1 |
| 3 | 120 | I | 1 |
| 7 | 144 | I | 1 |

**Table 5.2** Period it takes for each LED to hit the end of the recovery stage at 95% RH/85°C, as well as the likelihood that each unit belongs to the degradation mechanism. The units are numbered in order of increasing time to initial degradation [5].

| Unit # | TTF (time to failure) | Inlier (I)/Outlier (O) | Pr (FM 1) |
|---|---|---|---|
| 8 | 48 | I | 1 |
| 5 | 72 | I | 1 |
| 10 | 72 | I | 1 |
| 12 | 72 | I | 1 |
| 6 | 96 | I | 1 |
| 11 | 96 | I | 1 |
| 13 | 96 | I | 1 |
| 14 | 96 | I | 1 |
| 15 | 96 | I | 1 |
| 19 | 96 | I | 1 |
| 2 | 120 | I | 1 |
| 4 | 120 | I | 1 |
| 17 | 120 | I | 1 |
| 20 | 120 | I | 1 |
| 1 | 144 | I | 1 |
| 3 | 180 | I | 1 |
| 16 | 180 | I | 1 |
| 18 | 180 | I | 1 |
| 7 | 204 | I | 1 |
| 9 | 228 | I | 1 |



**Table 5.3** During the initial degradation and at the end of the recovery period, distribution parameters of each dominant degradation mechanism were measured at 95% RH/85°C [5].

| | Degradation conditions | |
|---|---|---|
| **Lognormal distribution parameters** | **During initial degradation** | **At the end of recovery stage** |
| $T_{50}$ (hours) | 30.75 | 49.98 |
| Shape parameters | 0.59 | 0.76 |
| Incubation time (hours) | 4.49 | 18.45 |

They explored the underlying physics for the deterioration in the second stages after exploring the number of potential failure mechanisms and their variation across the different stages of LED lifetime measured.

## 5.3 Analysis on the test samples

### 5.3.1 Modeling analysis for the testing

As shown in Fig. 5.5, they used mechanical ANSYS 14 Parametric Design Language (MAPDL) software to construct a 2D finite element model. In Fig. 5.3, the light blue color reflects silicone molding compound. The LED die was shown in red, and the die attach was shown in purple, sandwiched between the LED die and the copper heat sink. For the sake of simplicity, they skipped bond wires, believing that they would have little impact on the lumen degradation recovery mechanism.

After the meshing of the 2D geometry of LEDs, the material properties of each part were allocated. Moisture saturated concentration Csat, moisture absorption and desorption times, and moisture diffusivity are the three most important parameters for diffusion modeling.

The moisture diffusion process in a water permeable body can be described using an equation, according to them and others [10].

$$\partial C/\partial t = D\left(\partial^2 C / \partial x^2 + \partial^2 C / \partial y^2 + \partial^2 C / \partial z^2\right) \tag{5.1}$$

This equation is similar to the heat diffusion equation, so finite element modelling (FEM) of moisture diffusion was performed using the thermal diffusion model with the following model parameter changes as shown in Table 5.4, as well as Ma et al. [11].

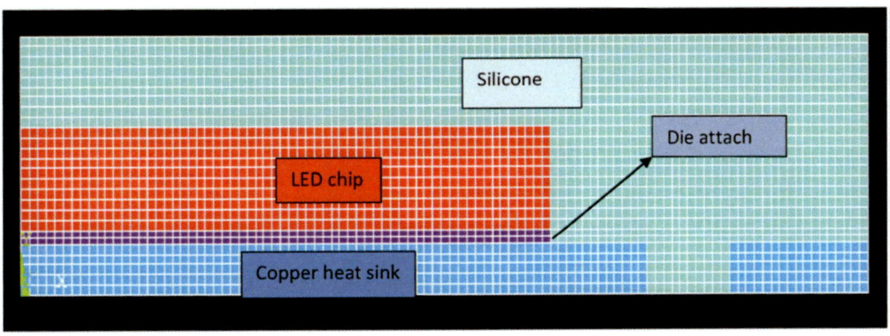


**Figure 5.3** ANSYS simulation using a 2D finite element model [5]. *(Reprinted with permission, © Microelectronics Reliability, Elsevier.)*

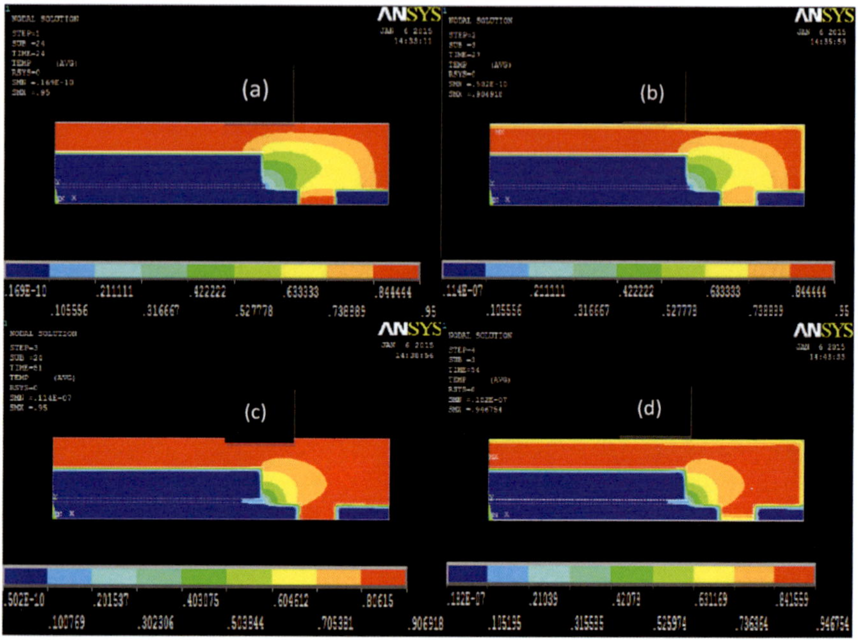


**Figure 5.4** ANSYS simulation of moisture distributions in the LED kit under 95% RH/85°C absorption and 60% RH/25°C desorption at (A) the end of the first absorption cycle (24 h); (B) the end of the first desorption period (27 h); (C) the end of the second absorption period (51 h); and (D) the end of the second desorption period (54 h) [5]. *(Reprinted with permission, © Microelectronics Reliability, Elsevier.)*

Csat denotes the highest moisture content that a material can absorb, while D denotes the moisture diffusivity of moisture in a material. Temperature, but not humidity, affects D [10]. W stands for normalized concentration or wetness, which was implemented to allow thermal modeling to be used for moisture modeling because, unlike temperature, moisture concentration does not have to be constant over an interface of two materials. W can be written in the following way [11]:

$$W = C/Csat^{*}T \tag{5.2}$$

where C is the moisture concentration ($kgm^{-3}$) in a material.

Moisture will reach the LED package while the samples are kept in the test chamber, and this duration is referred to as the moisture absorption period by them. Moisture evaporates to the ambient when the samples are taken out of the chamber every 24 h for characterization, and this duration is known as the moisture desorption period. Different diffusivity constants

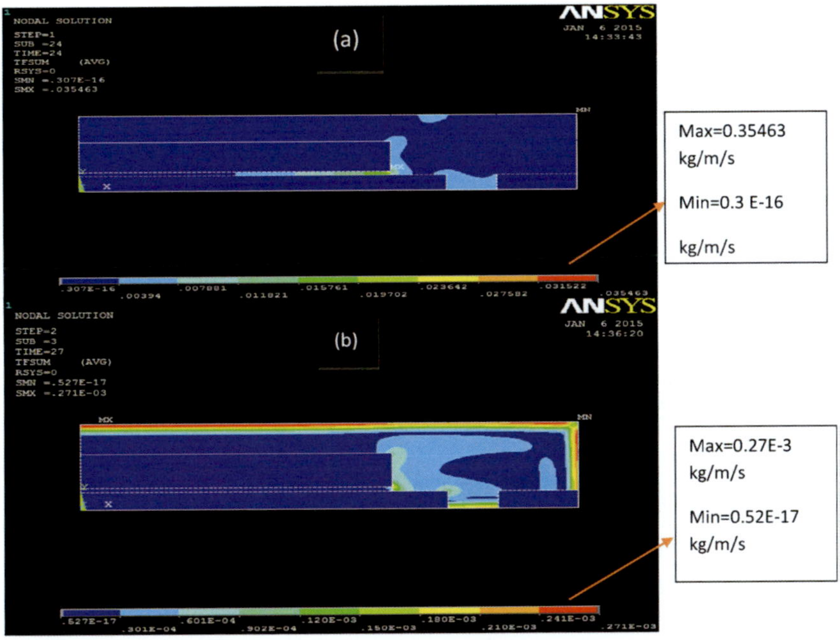


**Figure 5.5** ANSYS simulation of moisture flux distributions in the LED kit at (A) the end of the first absorption period (24 h) and (B) the end of the first desorption period (60 h) under 95% RH/85°C absorption and 60% RH/25°C desorption conditions (27 h) [5]. *(Reprinted with permission, © Microelectronics Reliability, Elsevier.)*

were needed to determine moisture diffusivity during absorption and desorption periods due to the different moisture transport mechanisms in these two periods.

Since copper heat sinks and LED dies do not absorb moisture, moisture conductivity (D* Csat) was set to a very low value of $1 \times 10^{-11}$ g/h × m for both, and their moisture specific capacity was taken as 1 g/$m^3$. Table 5.5 shows the other values that were obtained from the literature [10,12–15].

P was set to 0.95 for the test condition of 95% RH during FEM, and it was set to 0.6 for the ambient RH of 60% during their desorption case test. During absorption and desorption, the temperature was set to 85°C and 25°C, respectively.

## 5.4 Degradation mechanism

Fig. 5.4 shows the moisture distribution simulation effects in the LED kit during the absorption and desorption period (a–d cycle). Fig. 5.4 shows

**Table 5.4** Table of correspondences for the thermal/moisture analogy [5].

| Property | Thermal | Moisture |
|---|---|---|
| Primary variable | Temperature, T | Wetness. W |
| Density | ρ (kg/m$^3$) | 1 |
| Conductivity | k (W/m × °C) | D* Csat (kg/s × m) |
| Specific heat | c (J/kg × °C) | Csat (kg/m$^3$) |



**Table 5.5** Moisture modeling includes material parameters for various elements in LED packages [5].

| Element in LED package | 85% RH and 85 C [14,15] | | 95% RH and 85 C | | 60% RH and 25 C | |
|---|---|---|---|---|---|---|
| | D (m$^2$/h) | $C_{sat}$ (g/m$^3$) | D (m$^2$/h) | $C_{sat}$ (g/m$^3$) | D (m$^2$/h) | $C_{sat}$ (g/m$^3$) |
| Silicone encapsulant | 6.789E-9 | 375 | 6.78E-9 | 601 | 6.23E-10 | 98.09 |
| Die attach | 4.5E-8 | 6200 | 4.5E-8 | 6924 | 5.9E-22 | 4376 |
| LED dice | 1E-11 | 1 | 1E-11 | 1 | 1E-11 | 1 |
| Copper | 1E-11 | 1 | 1E-11 | 1 | 1E-11 | 1 |



how moisture enters the encapsulant during the absorption period and eventually reaches the die connect after a few cycles. Moisture evaporates from the silicone encapsulant during the desorption period since the moisture content in the encapsulant is greater than the ambient. With the absorption and desorption cycles, moisture continues to enter and reach toward the die attach, as shown in Fig. 5.5, which depicts the moisture flux distribution in the package.

Moisture flux was 0.35 kg/m/s during absorption and was located at the edge of the LED die and die attach interface, but it was $2.7 \times 10^{-4}$ kg/m/s during desorption, which was significantly lower than the absorption event. They came to the conclusion that moisture entering the die attach was much faster than moisture evaporating from the LED package through the silicone encapsulant. The findings that the amount of moisture evaporated during desorption was less than the amount absorbed, as shown in Fig. 5.6, and that the overall amount of moisture wetness in the package increased

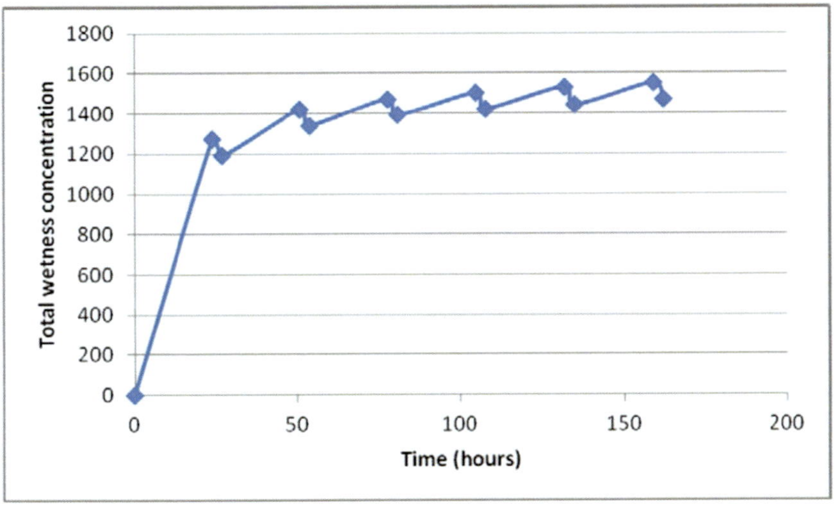


**Figure 5.6** In an LED package, total wetness content versus time. The absorption temperature is 95% RH/85°C, and the desorption temperature is 60% RH/25°C [5]. *(Reprinted with permission, © Microelectronics Reliability, Elsevier.)*

during a prolonged a–d period, as shown in Fig. 5.6, were found to be consistent.

According to the modeling results, a lot of moisture was absorbed in the first 24 h, resulting in light scattering and early light output depletion in LEDs [4]. Following that, the percentage increase in moisture absorption decreases during subsequent absorption periods. This was due to the fact that as more moisture accumulated in silicone, its concentration approaches Csat, thus reducing the amount of moisture that could be absorbed further. In fact, as shown in Fig. 5.6, the silicone surface reached its Csat between 50 and 100 h, forming a "safety layer" to resist more moisture absorption, and thus the percent increase in moisture plateaued as shown in Fig. 5.6.

Entrapped moisture would begin to diffuse to areas with lower moisture concentrations during the absorption process at a high temperature of 85°C, such as the die attach. Fig. 5.7 shows that about 78 h, more moisture was entering the die attach as opposed to the silicone encapsulant, where the moisture flux was now higher than that of the encapsulant. As a result, moisture in the encapsulant, especially near the LED chip, is sucked into the die attach, resulting in less moisture above the LED chip. On the other side, since the moisture layer on the surface of silicone prevents more moisture absorption, the amount of moisture in the region of silicone between its

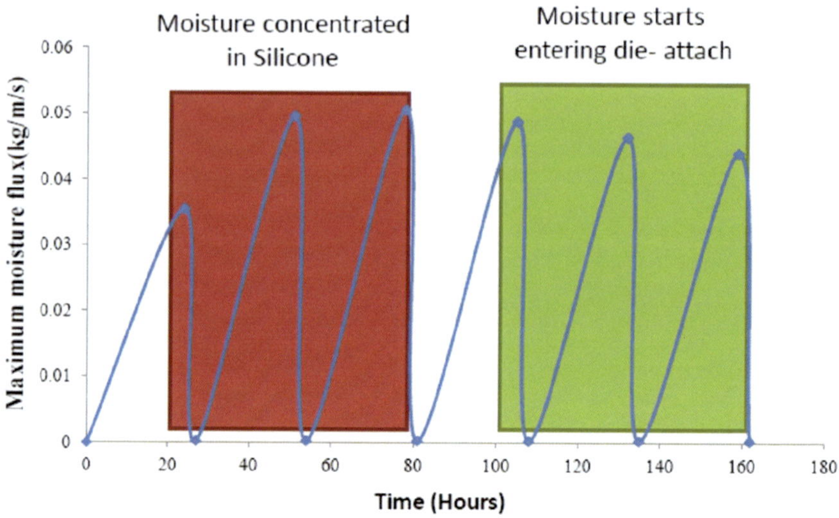


**Figure 5.7** Moisture flux variance simulated during humidity testing for LEDs at a humidity level of 95% RH/85°C [5]. *(Reprinted with permission, © Microelectronics Reliability, Elsevier.)*

surface layer and the die attach will decrease. After 78 h, this can result in an increase in lumen. This matches the experimentally observed lumen degradation recovery by Singh et al. who found that the lumen recovery time at 85°C and 95% RH was 72 h, which was similar to the simulation findings.

The safety layer at the silicone surface, on the other hand, collapsed as moisture from within the silicone diffused into the die attach. The role of Csat in protecting moisture diffusion was nullified as a result of this, allowing external moisture to diffuse continuously. The moisture content in the die attach, on the other hand, cannot be evaporated due to the die attach's deep location within the package. As a result, during the desorption stage, the moisture will remain within the package. Therefore, the recovery period was the start of the permanent deterioration stage, as Tan and Singh [3] discussed. The apparent recovery of lumen degradation was actually linked to a more severe degradation that occurs within the die attach. Hence, they demonstrated that lumen degradation recovery was not the same as reliability rejuvenation.

They also discovered that, as shown in Fig. 5.8, the average moisture content of silicone decreases as more moisture reaches the die attach. They discovered that this occurred at 78 h as well, and they conclude that the

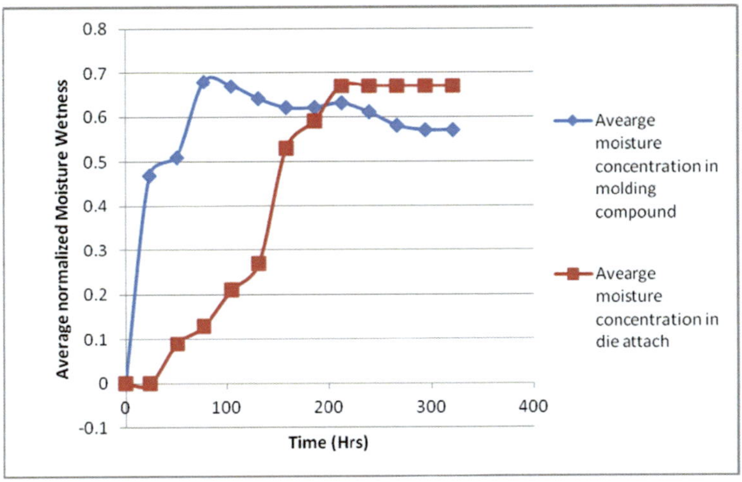


**Figure 5.8** After absorption, average moisture wetness in silicone encapsulation and die attach [5]. *(Reprinted with permission, © Microelectronics Reliability, Elsevier.)*

decrease in moisture content in the encapsulant was the cause of the lumen degradation recovery seen in their study.

The moisture solubility in die attachment induced the saturation of the average moisture concentration, as shown in Fig. 5.8. More moisture will enter the die attach, soaking more die attachment amount, and due to its location between the heat sink and the LED dice, die attachment will have no way to release the moisture. As moisture accumulates in the die attachment, hygro-swelling tension develops, eventually leading to die attachment delamination.

Singh et al. [5] calculated the time when maximum moisture flux shifted from silicone encapsulant to die attachment at both moisture levels to check

**Table 5.6** At 85% RH and 95% RH, the time for peak lumen degradation and the time for maximum moisture flux transfer from silicone to die connect are compared, with both temperatures being 85°C [5].

| 85% RH and 85°C | | 85% RH and 85°C | |
|---|---|---|---|
| Time for peak lumen degradation | Time for moisture flux shift from silicone to die attachment | Time for peak lumen degradation | Time for moisture flux shift from silicone to die attachment |
| 120 h | 105 h | 72 h | 78 h |

this process. The results of the simulation closely agree with the experimental comparisons as shown in Table 5.6. From 95% RH to 85% RH, the computed peak average moisture concentration ratio in silicone was 1.12, which was found to be equivalent to the experimental ratio of peak lumen degradation from 95% RH to 85% RH of 1.25.

### 5.4.1 Experimental verification of the proposed mechanism

Singh et al. [5] used C-SAM to analyze the LED dice during the initial rapid deterioration period and recovery stage to further validate the proposed mechanisms. After a test period of 48 h (corresponding to the initial degradation) and 550 h (corresponding to the recovery phase), LEDs were chosen at random for investigation.

For the C-SAM examination, LEDs were flipped upside down (i.e., the lens was facing downward) as shown in Fig. 5.9. The depth of the return echoes was gated between the backside mold compound and the backside of the die paddle, and the interface between the chip and the die attachment material. Tiny voids (trapped bubbles) at the mold compound/die paddle interface and the LED chip/die attachment interface are visible in the acoustic images. In Fig. 5.9, the die attachment region is easily defined by a square positioned exactly in the center of the LED box.

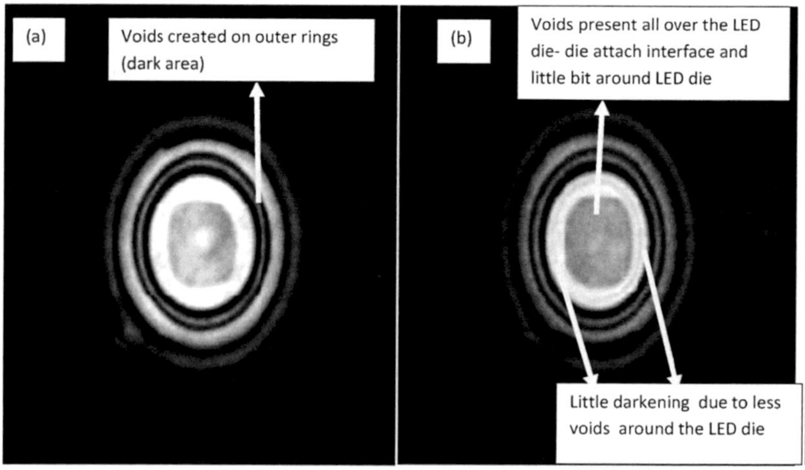


**Figure 5.9** C-SAM of a flipped LED (A) before (tested for 48 h) and (B) after initial degradation (tested for 550 h). After 550 h of testing, the die connect interface with the LED die was completely blackened, but the area outside the LED die was not as darkened [5]. *(Reprinted with permission, © Microelectronics Reliability, Elsevier.)*

Fig. 5.9A shows the C-SAM results for the LEDs tested for 48 h, which showed void (dark area) at the outer rings of the LEDs, indicating that moisture was only present in the molding compound. Fig. 5.9B shows that for LEDs tested until 550 h, and the entire die attach area was filled with black spots, indicating that moisture had reached the die attach underneath the LED die and produced voids, which could lead to LED die attach delamination. These findings matched the ANSYS simulation results shown in Fig. 5.8.

When comparing the LEDs tested for 550 h (also seen in Fig. 5.9B) to the LEDs tested for 48 h, Singh et al. [5] discovered that the area around the LED die (molding compound) was only slightly darkened. This suggests that the LEDs tested for 550 h accumulated less moisture in the molding compound than those tested for 48 h, likely due to moisture sucking from the molding compound to the die attach for the extended testing, as shown by their finite element analysis.

The presence of moisture was shown in both cases by a minor darkening of the molding compound, which explained why LEDs did not completely recover to 0% lumen degradation after the recovery process.

They also claimed that chip level failure is unlikely due to the two-stage degradation phase that occurred during the early life of LEDs. The results of their parameter extraction technique are shown in Table 5.7.

They discovered that the values of n, Is, and Rs did not change in Table 5.7, suggesting that no chip or bonding wire degradation had occurred. Apart from the interface cracks shown in Fig. 5.9, there were no defects found on the device. As a result, they were able to confirm that moisture was the sole cause of lumen degradation, which was also supported by simulation results.

**Table 5.7** The n, Rs, and Is values for the LEDs tested at 95% RH/85°C during the initial degradation and at the end of the recovery stage are summarized in the table below [5].

| Parameters | During initial degradation | At the end of recovery stage |
|---|---|---|
| n – ideality factor | 2.74 | 2.78 |
| Is (A) – reverse saturation current | $4.66 \times 10^{-19}$ | $5.39 \times 10^{-19}$ |
| Rs (ϙ) – series resistance | 0.16 | 0.19 |

## 5.5 Implication to temperature–humidity test for LEDs

According to the findings of the above report, permanent degradation starts at 78 h for the 95% RH case test condition with a 24 h absorption interval and a 3 h desorption interval. Similar simulations were carried out by Singh et al. [5], but with different absorption and desorption time intervals. Table 5.8 shows the permanent degradation start time under various conditions, and it can be shown that degradation recovery is faster for longer absorption times and smaller desorption times, as predicted. This also demonstrates that desorption periods have a greater effect than absorption time intervals.

Singh et al. [5] found that if the absorption was performed continuously, permanent degradation would start at the shortest time, which is 55.4 h. If the absorption time is reduced to 24 h, and the desorption time is reduced to 3 h, the time to achieve the full moisture flux is 78 h. As a result, they discovered that different characterization time intervals during the temperature–humidity test would affect the permanent degradation's starting time, and thus the recorded reliability, even under the same test conditions. To put it another way, in order to provide reliable humidity reliability data for LEDs, the humidity test protocol for LEDs needs to be standardized in terms of the characterization interval.

**Table 5.8** Variation in the onset of irreversible degradation for high-power LEDs with varying absorption and desorption periods at 95% RH and 85°C absorption and 60% RH and 25°C desorption [5].

| ptTesting procedure | Time to reach moisture recovery in silicone i.e., moisture flux = 0.05 kg/m/s (hours) |
|---|---|
| Continuous absorption | 55.4 |
| Absorption for 24 h and then desorption for 1 h | 57 |
| Absorption for 24 h and then desorption for 3 h | 78 |
| Absorption for 48 h and then desorption for 1 h | 56.7 |
| Absorption for 48 h and then desorption for 3 h | 61 |

## 5.6 Summary

The investigation of the lumen degradation recovery mechanism after a rapid degradation mechanism was the subject of this chapter. To understand the physics underlying the lumen degradation recovery phenomenon, a 2D finite element model was made and moisture absorption—desorption simulations were run using ANSYS software. Moisture accumulation under the LED die and die attach delamination due to moisture were verified using C-SAM results. The recovery was discovered to be due to moisture sucking into the die attach, which resulted in lower moisture in the silicone encapsulant, according to both simulation and experimental findings. As a result, this reversal of lumen degradation is simply the start of a long-term degradation stage.

## References


[1] Y.H. Lin, J.P. You, Y.-C. Lin, N.T. Tran, F.G. Shi, Development of high-performance optical silicone for the packaging of high-power LEDs, IEEE Transactions on Components Packaging and Technology 33 (4) (2010) 761—766.

[2] Z. Ma, X. Zheng, W. Liu, X. Lin, W. Deng, Fast thermal resistance measurement of high brightness LED, in: Proceedings of the 6th International Conference Electronic Packaging Technology, September, 2005, pp. 1—3.

[3] C.M. Tan, P. Singh, Time evolution degradation physics in high power white LEDs under high temperature—humidity conditions, IEEE Transactions on Device and Materials Reliability 14 (June, 2014) 742—750.

[4] C.M. Tan, B.K. Chen, X. Li, S.H. Chen, Rapid light output degradation of GaN-based packaged LED in the early stage of humidity test, IEEE Transactions on Device and Materials Reliability 12 (1) (March, 2012) 44—48.

[5] P. Singh, C.M. Tan, L.B. Chang, Early degradation of high power packaged LEDs under humid conditions and its recovery—myth of reliability rejuvenation, Microelectronics Reliability 61 (2016) 145—153.

[6] C.M. Tan, P. Singh, Extrapolation of Lifetime of High Power LEDs under Temperature—Humidity Conditions, EDSSC, 2014.

[7] C.M. Tan, N. Raghavan, Simulated Annealing for Mixture Distribution Analysis and its Applications to Reliability Testing, INTECH Open Access Publisher, 2008.

[8] C.M. Tan, N. Raghavan, An approach to statistical analysis of gate oxide breakdown mechanisms, Microelectronics Reliability 47 (9) (2007) 1336—1342.

[9] N. Raghavan, C.M. Tan, Statistical analysis of multi-censored electromigration data using the EM algorithm, Physical and Failure Analysis of Integrated Circuits, in: 2007. IPFA 2007. 14th International Symposium on the IEEE, 2007.

[10] E.H. Wong, Y.C. Teo, T.B. Lim, Moisture diffusion and vapor pressure modeling of IC packaging, in: Proceeding of the 48th Electronic Computing Technology Conference, 1998, pp. 1372—1378.

[11] X.S. Ma, K.M.B. Jansen, L.J. Ernst, W.D. van Driel, O. van der Sluis, G.Q. Zhang, C. Regard, C. Gautier, H. Frémont, Moisture diffusion model verification of packaging materials, in: K.Y. Bi, F. Xiao (Eds.), Proceedings of International Conference on Electronic Packaging Technology & High Density Packaging, 2008, pp. 380—384.

[12] E.H. Wong, K.C. Chan, T.Y. Tee, R. Rajoo, Comprehensive treatment of moisture induced failure in IC packaging, in: Proceedings of the 3rd International Electronic Manufacturing Technology, Japan, 1999, pp. 176–181.
[13] T.F. Lam, FEA simulation on moisture absorption in PBGA package under various moisture pre-conditioning, in: Proceedings of the 50th Electronic Computing Technology Conference, 2000, pp. 1078–1082.
[14] T.Y. Tee, X.J. Fan, T.B. Lim, Modeling of Whole Field Vapor Pressure during Reflow for Flip Chip BGA and Wire Bond PBGA Packages, Proc. 1st Electron. Mater. Packag. Workshop, Singapore, 1999, pp. 38–45.
[15] R.L. Shook, D.L. Gerlach, B.T. Vaccaro, Moisture blocking planes and their effect on reflow performance in achieving reliable Pb-free assembly capability for PBGAs, in: Proceedings of the 51st Electronic Computing Technology Conference, 2001, pp. 74–79.

CHAPTER 6

# Permanent degradation mechanisms of high-power packaged white LEDs

## 6.1 Introduction

According to Singh and Tan's work [1], the structure of LEDs differs significantly from that of standard integrated circuits, and the final degradation pattern due to wearout can vary from that of integrated circuits. This chapter will concentrate on the final stage of degradation, and the experiments presented are based on Tan and Singh's work [2].

They put 60 high-power OSRAM white packaged LEDs with 1 W power through an accelerated test in a high humidity—temperature chamber. Three humidity levels were investigated, each with a sample size of 20 samples, and all three cases were held at temperature of 85°C. The three RH levels used were 70, 85, and 95% relative humidity.

Within the test chamber, LEDs were fixed on a metal board and positioned. To avoid the evaporation of moisture trapped in the LED's encapsulant, LEDs were tested in an OFF state, meaning no power was applied during the test. After a fixed period of 24 h, LEDs were removed from the test chamber to assess their electrical and optical properties. To ensure that no moisture from the LED package was evaporated, all measurements were completed within 3 h.

Using a Keithley source meter 2651A and an integrating sphere, the electrical I-V and optical characteristics of each LED, such as luminous flux and blue-to-yellow ratio (BYR), were measured. Fig. 6.1 depicts the optical measurement device they used in their research. It consists of an integrating sphere of 1 m diameter and model number SLM-40TS-110,902 and an Ocean Optics spectroradiometer model number QE6500. For each LED, the initial measurements were used as a reference baseline.

*Reliability and Failure Analysis of High-Power LED Packaging*
ISBN 978-0-12-822408-3
https://doi.org/10.1016/B978-0-12-822408-3.00009-5

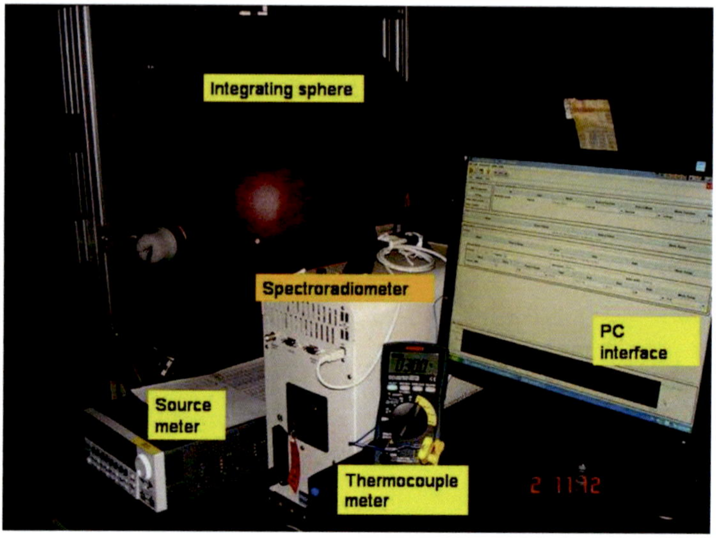


**Figure 6.1** Experiment's optical measurement system [2]. *(IEEE Transactions on Device and Materials Reliability, IEEE. Reprinted with permission.)*

A steady current of 300 mA was used to power the LEDs during measurements to ensure measurement accuracy. They used a thermoelectric cooler to keep the ambient temperature at 30°C and kept the stabilization time at 30 s to reduce measurement noise.

The degradation time to different % lumen degradation in comparison to their initial value was calculated using the measured luminous fluxes of all the LEDs plotted versus time. The results for the three sets of LEDs under different humidity at the same temperature conditions are shown in Table 6.1.

## 6.2 Statistical analysis on the test results

### 6.2.1 Identification of degradation mechanisms-related statistical distributions

They used the Akaike information criterion (AIC), as defined in Chapters 4 and 5, to estimate the number of different degradation mechanisms that could exist in the given collection of data. In the test case under 95% RH, at each percent deterioration, they revealed three probable degradation processes, as represented in Tables 6.2–6.4. The corresponding statistical distributions, as determined by Simulated Annealing (SA) and Expectation and Maximization (EM), were three-parameter lognormal distributions.

**Table 6.1** At 85°C and three humidity levels, the time it takes for each LED to degrade is measured in hours [2].

| | At 95% RH | | | At 85% RH | | | At 70% RH | | |
|---|---|---|---|---|---|---|---|---|---|
| | Lumen degradation % | | | Lumen degradation % | | | Lumen degradation % | | |
| Sample # | 5% | 10% | 15% | 5% | 10% | 15% | 5% | 10% | 15% |
| 1 | 65 | 88 | 116 | 485 | 700 | 890 | 91.65 | 128.9 | 184.6 |
| 2 | 39 | 42 | 61 | 360 | 600 | 720 | 150 | 212 | 215 |
| 3 | 180 | 362 | 535 | 240 | 250 | 260 | 42 | 144 | 315 |
| 4 | 500 | 608 | 639 | 290 | 400 | 525 | 70 | 145 | ND |
| 5 | 310 | 505 | 580 | 30 | 150 | 340 | 83 | 140 | 191 |
| 6 | 21 | 43 | 65 | 80 | 91 | ND | 90 | 115 | 140 |
| 7 | No degradation (ND) | | | 720 | 820 | 925 | 100 | 140 | 155 |
| 8 | | | | 270 | 480 | 680 | 160 | 170 | 190 |
| 9 | | | | 100 | ND | | 76 | 100 | 125 |
| 10 | 110 | 238 | 560 | 710 | 780 | 855 | 123 | 189 | 240 |
| 11 | 95 | 510 | 690 | 415 | 445 | 480 | 46 | 141 | 242 |
| 12 | 20 | 35 | 60 | 150 | 160 | 165 | 13 | 45 | 110 |
| 13 | 23 | 37 | 300 | 270 | 340 | ND | 20 | 53 | 115 |
| 14 | 140 | 228 | 540 | 95 | 105 | ND | 140 | 180 | 200 |
| 15 | 400 | 440 | 482 | 270 | 340 | 400 | 110 | 115 | 150 |
| 16 | 430 | 440 | 450 | 135 | ND | | 30 | 105 | 180 |
| 17 | 200 | 222 | 250 | 760 | 900 | ND | 150 | ND | |
| 18 | 23 | 100 | 355 | 170 | 220 | 280 | 97 | 200 | 290 |
| 19 | 30 | 55 | 100 | 150 | 220 | 275 | 100 | 194 | 300 |
| 20 | 45 | 424 | 510 | 200 | 290 | 400 | 165 | 210 | ND |

IEEE Transactions on Device and Materials Reliability, IEEE. Reprinted with permission.

**Table 6.2** Each LED's time to 5% degradation under 95% relative humidity/85°C, as well as the probability (Pr) of each potential statistical distribution associated with each degradation process (DM). The units 3 and 12 are outliers [2].

| Unit # | TTD (hour) | Pr (DM 1) | Pr (DM 2) | Pr (DM 3) |
|---|---|---|---|---|
| 12 | 20 | 0 | 0 | 0.07 |
| 6 | 21 | 0 | 0.92 | 0.07 |
| 13 | 23 | 0 | 0.92 | 0.07 |
| 18 | 23 | 0 | 0.92 | 0.07 |
| 19 | 30 | 0 | 0.92 | 0.07 |
| 2 | 39 | 0 | 0.92 | 0.07 |
| 20 | 45 | 0 | 0.92 | 0.07 |
| 1 | 65 | −5.00E-280 | 0.92 | 0.07 |
| 11 | 95 | −2.00E-104 | 0.92 | 0.07 |
| 10 | 110 | −4.80E-20 | 0.93 | 0.06 |
| 14 | 140 | −9.10E-06 | 0.94 | 0.05 |
| 3 | 180 | 2.13 | −1.07 | −0.05 |
| 17 | 200 | 7.46E−17 | 1.00 | −0.01 |
| 5 | 310 | 2.00E−115 | 0.74 | 0.25 |
| 15 | 400 | 9.40E−191 | 0.81 | 0.18 |
| 16 | 430 | 2.40E−215 | 0.82 | 0.17 |
| 4 | 500 | 3.80E−271 | 0.83 | 0.16 |



**Table 6.3** Time to 10% degradation of each LED at 95% relative humidity/85°C, as well as its respective Pr to each potential statistical distribution connected with each dm. Units 3 and 12 are outliers [2].

| Unit # | TTD (hour) | Pr (DM 1) | Pr (DM 2) | Pr (DM 3) |
|---|---|---|---|---|
| 12 | 35 | −0.01 | 1.00 | 5.11E-05 |
| 13 | 37 | 0.93 | 0.06 | −0.01 |
| 2 | 42 | 0.94 | 0.05 | −0.01 |
| 6 | 43 | 0.95 | 0.04 | −0.01 |
| 19 | 55 | 0.95 | 0.04 | −0.01 |
| 1 | 88 | 0.97 | 0.03 | −0.01 |
| 18 | 100 | 0.98 | 0.02 | −0.01 |
| 17 | 222 | 0.98 | 0.02 | −0.01 |
| 14 | 228 | 0.99 | 0.02 | −0.02 |
| 10 | 238 | 1.00 | 0.02 | −0.02 |
| 3 | 362 | 1.49 | 0.03 | −0.53 |
| 20 | 424 | 0.84 | 0.02 | 0.12 |
| 15 | 440 | 0.86 | 0.02 | 0.10 |
| 16 | 440 | 0.86 | 0.02 | 0.10 |
| 5 | 505 | 0.89 | 0.03 | 0.07 |
| 11 | 510 | 0.89 | 0.03 | 0.07 |
| 14 | 608 | 0.90 | 0.03 | 0.05 |

**Table 6.4** Units 16 and 17 are outliers in terms of time to 15% deterioration of each LED under 95% RH/85 C and its respective Pr to each feasible statistical distribution connected with each DM is shown [2].

| Unit # | TTD (hour) | Pr (DM 1) | Pr (DM 2) | Pr (DM 3) |
|---|---|---|---|---|
| 12 | 60 | 1.00 | −0.01 | −0.01 |
| 2 | 61 | 1.00 | −0.01 | −0.01 |
| 6 | 65 | 1.00 | −0.01 | −0.01 |
| 19 | 100 | 1.00 | −0.01 | −0.01 |
| 1 | 116 | 1.01 | −0.01 | −0.01 |
| 17 | 250 | 0.23 | 0.76 | −0.01 |
| 13 | 300 | 1.01 | 0.01 | −0.01 |
| 18 | 355 | 1.02 | 0.01 | −0.03 |
| 16 | 450 | −0.08 | −0.01 | 1.08 |
| 15 | 482 | 0.88 | 0.01 | 0.11 |
| 20 | 510 | 0.92 | 0.01 | 0.07 |
| 3 | 535 | 0.94 | 0.01 | 0.05 |
| 14 | 540 | 0.94 | 0.01 | 0.05 |
| 10 | 560 | 0.95 | 0.01 | 0.04 |
| 5 | 580 | 0.95 | 0.01 | 0.04 |
| 4 | 639 | 0.96 | 0.01 | 0.03 |
| 11 | 690 | 0.96 | 0.01 | 0.02 |



While AIC established three potential degradation mechanisms, there were a few outlier points, and after a thorough review of Tables 6.2–6.4, they determined that each percent degradation level had only one dominant degradation mechanism. Table 6.5 summarizes degradation parameters at each percent degradation step, and Fig. 6.2 depicts the related lognormal likelihood plots. To plot a three-parameter log-normal distribution on a two-parameter normal log probability graph paper, the degradation times are subtracted by the corresponding incubation times [3].

**Table 6.5** Distribution parameters are shown for each dominant degradation mechanism at relative humidity 95%/85°C [2].

| | Degradation % | | |
|---|---|---|---|
| **Lognormal distribution parameters** | **5%** | **10%** | **15%** |
| $T_{50}$ (hours) | 99.13 | 147.21 | 186.72 |
| Shape parameter | 0.5 | 0.26 | 0.34 |
| Incubation time (hours) | 0 | 0 | 0 |

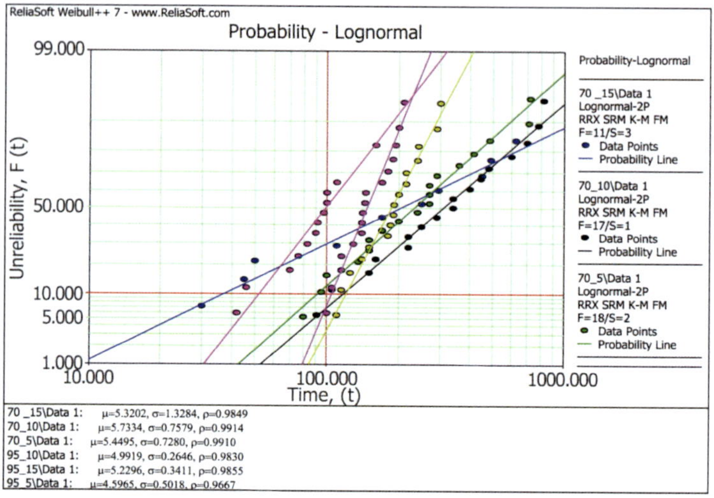


**Figure 6.2** At 95% and 70% RH, time to deterioration is plotted as a lognormal likelihood plot at 85°C. To prevent congested lines, 85% RH data are not included. The legend mark 70 15 denotes 70% RH and 15% degradation, and the same is true for the other legend labels [2]. *(IEEE Transactions on Device and Materials Reliability, IEEE. Reprinted with permission.)*

For the test results under 70 and 85% RH, they also discovered one dominating failure cause at each percent deterioration stage, and their distribution parameters are reported in Tables 6.6 and 6.7, respectively.

The shape parameter at each percent degradation was found to be different, as seen in Tables 6.5 and 6.6, indicating that the degradation mechanism at each percent degradation was different, and thus, the underlying physics was evolving over time [3,4]. Furthermore, the shape parameters vary at a given percent degradation for different percent RH levels, indicating that extrapolation from 70% to 85% to 95% RH levels is

**Table 6.6** At 85% relative humidity and 85°C, the distribution parameters of each dominant degradation pathway were determined [2].

| | Degradation % | | |
|---|---|---|---|
| **Lognormal distribution parameters** | **5%** | **10%** | **15%** |
| $T_{50}$ (hours) | 110 | 185 | 403.43 |
| Shape parameter | 1.97 | 2.89 | 0.43 |
| Incubation time (hours) | 19 | 30 | 60 |

IEEE Transactions on Device and Materials Reliability, IEEE. Reprinted with permission.

**Table 6.7** At 70% RH and 85°C, the distribution parameters of each main degradation mechanism were determined [2].

| | Degradation % | | |
|---|---|---|---|
| **Lognormal distribution parameters** | **5%** | **10%** | **15%** |
| $T_{50}$ (hours) | 232.64 | 389.02 | 434.44 |
| Shape parameter | 0.73 | 0.75 | 1.33 |
| Incubation time (hours) | 0 | 0 | 230 |

IEEE Transactions on Device and Materials Reliability, IEEE. Reprinted with permission.

not feasible. Since the acceleration was not linear, test data at 85°C and 85% RH cannot be used to predict LED lifetime under lower humidity conditions. Readers may look up the definition of linear acceleration in Refs. [3,4].

In the test done under 70% RH, however, shape characteristics at 5% and 10% degradation levels were identical, implying that a common degradation mechanism exists up to 10% degradation and that a new degradation mechanism emerges after that. According to Tan et al. [2], the deterioration process and pattern would be more important in this test condition than in the other two because it is closer to typical operating condition.

## 6.3 Physical analysis on the test samples

### 6.3.1 Identification of the degradation physics

The underlying degradation physics of these LEDs were investigated after determining the number of possible degradation mechanisms for each collection of LEDs. Degradation of a packaged LED may occur at either the package or chip level. At the package level, bonding thread, bond pad, and encapsulant (silicone gel and lens) degradation can be considered separately, while at the chip level, LED dice and phosphor degradation can be considered separately.

### 6.3.2 Nondestructive analysis

To study chip-level deterioration, BYR, as well as ideality factor and reverse saturation current of the LED junction, was investigated. By dividing the peak intensity at blue wavelength by the peak intensity at yellow wavelength, BYR was calculated. The transition in BYR for two units from the test at 95% RH is depicted in Fig. 6.3 at various stages of

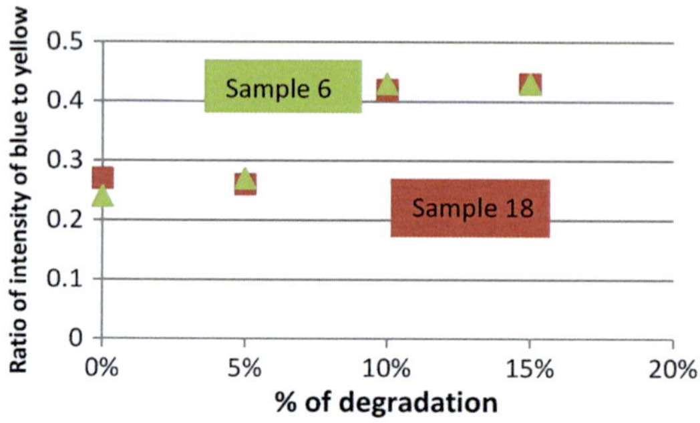


**Figure 6.3** Under stress circumstances of 95% RH and 85°C, the blue-to-yellow spectrum strength ratio of radiated light versus percent of degradation for two chosen units [2]. *(IEEE Transactions on Device and Materials Reliability, IEEE. Reprinted with permission.)*

degradation. These units were chosen because they had a greater change in BYR values, which made the degradation physics more apparent, and the other units in the set were supposed to follow a similar pattern because they all came from the same statistical distribution.

The ideality factor and reverse saturation current for all LEDs remained consistent up to 15% deterioration. These numbers were derived using the parameter extraction method described in previous chapters. Junctions were found to have not degraded at different percent of degradation as there was no noticeable difference in blue light peak wavelength, ideality factor, or reverse saturation current for all LEDs.

According to the extraction results, the series resistance (Rs) of the units changes with the degradation percent level and then remains constant above the 5% degradation level. The variation of the two samples chosen is depicted in Fig. 6.4.

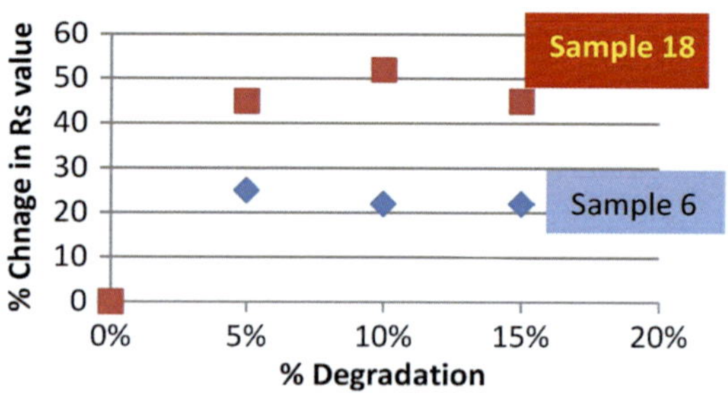


**Figure 6.4** Changes in Rs for the two units depicted in Fig. 6.3 at various stages of degradation under stress conditions of 95% relative humidity and 85°C [2]. *(IEEE Transactions on Device and Materials Reliability, IEEE. Reprinted with permission.)*

Therefore, the first 5% of degradation (i.e., 0%—5% degradation) was caused by an increase in Rs, and the following 5% (i.e., 5%—10% degradation) was caused by phosphor degradation, as evidenced by an increase in BYR when the LED chip was confirmed to be intact (see Figs. 6.3 and 6.4). Another process was responsible for the final 5%. These results were consistent with the statistical study, which showed that the degradation process was changing over time. Tan et al. [2] stated that similar data analysis on LED sets tested at 85% RH and 85°C will be insignificant because their process does not resemble the usual operating conditions for LEDs; instead, they move to the analysis of the test results at 70% RH and 85°C.

The ideality factor and Rs of all units exposed to 70% RH and 85°C remained unchanged, as did the blue light's peak wavelength, suggesting that the LED dices were unharmed. The BYR pattern of the two typical units is shown in Fig. 6.5, which shows a minor increase of less than 10% during the early degradation. Because the shape characteristics of the degradation distribution do not vary for deterioration up to 10%, a change in BYR was deemed to be insignificant, and it was not the source of degradation. However, as shown in Fig. 6.6, the increase in Rs was substantial and consistent, leading to the conclusion that the dominant degradation mechanism is the increase in Rs at this stress level.

Tan et al. [2] also concluded that from 10% degradation and onward, there may be another degradation mechanism since the Rs rise from 10% to

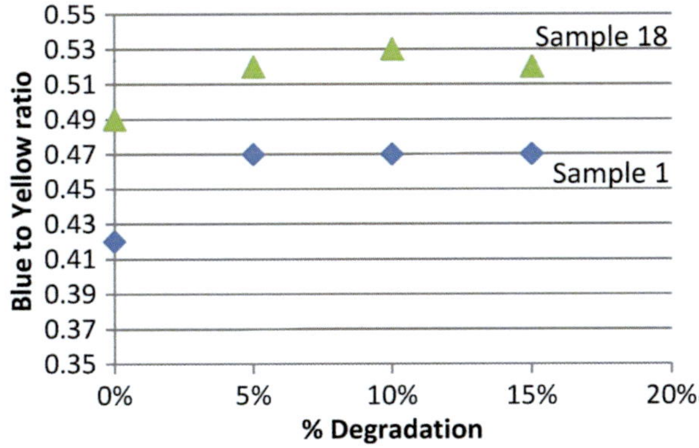


**Figure 6.5** At a stress situation of 70% RH and 85°C, the blue-to-yellow spectrum intensity ratio of the radiated light versus percent deterioration [2]. *(IEEE Transactions on Device and Materials Reliability, IEEE. Reprinted with permission.)*

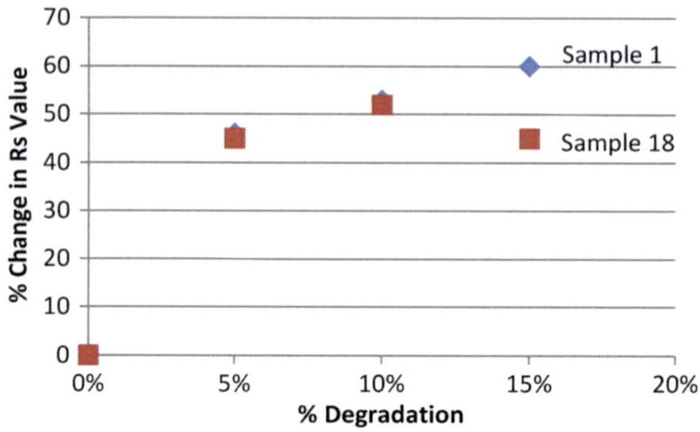


**Figure 6.6** Changes in the Rs of the two selected units at varying levels of deterioration at 70% RH and 85°C are illustrated in Fig. 6.5 [2]. *(IEEE Transactions on Device and Materials Reliability, IEEE. Reprinted with permission.)*

15% degradation was becoming very small. As shown in Table 6.7, this was also found to be in accordance with the statistical analysis.

## 6.4 Degradation mechanism

### 6.4.1 Physical destructive analysis

The possible and probable causes of degradation mechanisms at different percent RH levels and different percent degradation levels at each percent RH level were investigated after the nondestructive study. Fig. 6.7 shows a

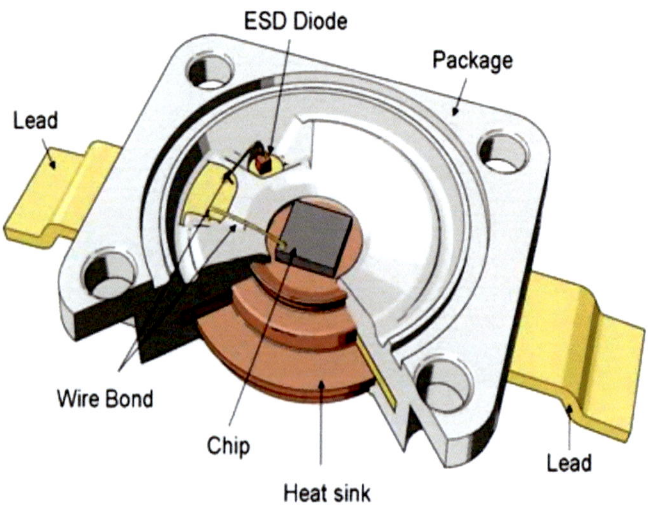


**Figure 6.7** High-power LEDs investigated in their work [2]. *(IEEE Transactions on Device and Materials Reliability, IEEE. Reprinted with permission.)*

closer look at the internal structure of a packaged LED [5] that they used in their research.

The bond pad used in LEDs, as shown in Fig. 6.7, was coated with Ag and acts as a silver mirror, as determined by the Energy Dispersive System (EDS) study. The Au bonding wire was adhered to the Ag coating directly. Kirkendall voids can easily form at the interface between Au and Ag because Ag is a quick diffuser in Au and Au is a sluggish diffuser in Ag [6,7]. At higher temperatures, such as 85°C, as utilized by Tan et al., this Kirkendall void formations will speed up, and when Ag is diffused into Au, it will be totally diffused into Au because there is a perfect solid solubility between Ag-Au binary system [8].

Moisture is also expected to penetrate the LED package via the encapsulant, wire—package interface, and lens—package interface due to the high temperature—humidity conditions used in the tests [5]. As mentioned in Chapter 5 and elsewhere [9], this has already been verified. As a result, at the Ag-Au interaction, water would surely be present. Since Au has a far higher electrochemical potential than Ag (Au is 1.52 V and Ag is 0.8 V at 25°C) [10], galvanic corrosion will occur at Au-Ag contacts on the bond pad. Ag will serve as an anode in a high temperature—humidity test, and anodic dissolution will occur, resulting in a removal rate of up to 35.6 nm/s [11]. After dissolution, Ag migrates via ion transport and electrodeposition. Electrodeposition results in Ag clouds or dendrites, and the activation energy of $Ag^+$ diffusion in water has been determined to be 0.2 eV [12]. After entering the LED package, moisture will travel via the gap along the peripheral of the Au ball to the ball—Ag coating contact in the bond pad construction [2].

Tan et al. [2] discovered that all of the above can be observed under SEM. Fig. 6.8 shows an SEM micrograph of Ag coating around the balls region in the LED package for both the fresh and deteriorated samples, which were stressed for 184.6 h at 85°C and 70% RH. After decapsulation, the Au bonding balls were displaced with an air gun to reveal the underlying region underneath the ball. Because the bond pad's encapsulant polymer layer cannot be removed during decapping, the bond pad for damaged LEDs cannot be seen under 95% RH and 85°C. At high temperatures, the composition of the encapsulant polymer substance changes, this was also a frequent occurrence during failure analysis by others [13,14]. In reality, if the heat is high enough, this change can happen in as little as 0.1 s [15].

The Rs increase as observed in Tan et al.'s work [2] was unlikely due to LED dice deterioration since they confirmed that the electrical parameters

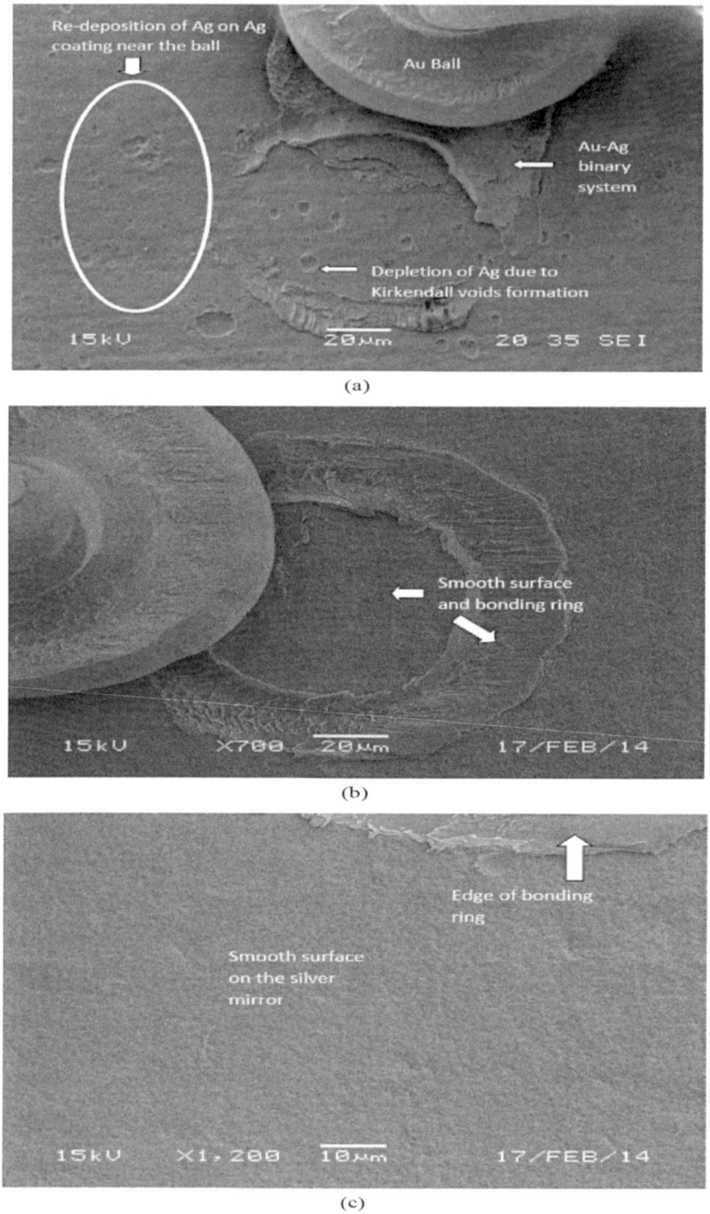


**Figure 6.8** SEM micrograph of the silver coating near the gold bond ball of (A) a degraded LED (unit #1) after 184.6 h at 70% RH and 85°C; (B) a new LED; (C) the corresponding rounded area of (A) on the new unit [2]. *(IEEE Transactions on Device and Materials Reliability, IEEE. Reprinted with permission.)*

do not change in any of the test cases, and such bond weakening was predicted to become more extreme at higher humidity levels such as 95% RH in Tan et al.'s work [2]. In fact, they discovered that in the 95% RH case, the polymer shift on top of the bond pad was more significant than in the 70% RH case. This may explain why, in the case of 95% RH, higher Rs and thus higher Joule heating resulted in polymer material change on the top.

Additional contaminated layers were discovered on the Ag coating, possibly due to a change in the polymer material composition [16,17]. Figs. 6.9 and 6.10 indicate this for the same unit used in Fig. 6.8. Only at 5 KV did some very thin layers become visible, as shown in Figs. 6.9 and 6.10. Figs. 6.11 and 6.12 display EDS spectra at 5 and 15 KV for a contamination layer spot indicated in Fig. 6.9, this layer was on top of the Ag coating, proving its existence.

The excessive bond pad resistance Rs rises and ceases after 5% degradation for units tested at 95% RH, and phosphor degradation becomes dominant after that, as shown in Fig. 6.3. They hypothesized that after this 5% degradation stage, the Ag-Au ball contact's joule heating increased the

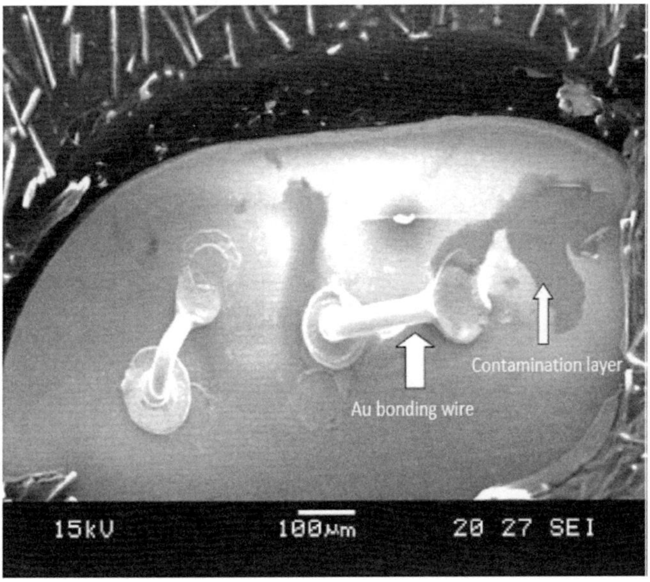


**Figure 6.9** The contaminated layer is visible on the LED's Ag-coated bond pad. The damage caused by decapsulation resulted in the broken Au bonding wire [2]. *(IEEE Transactions on Device and Materials Reliability, IEEE. Reprinted with permission.)*

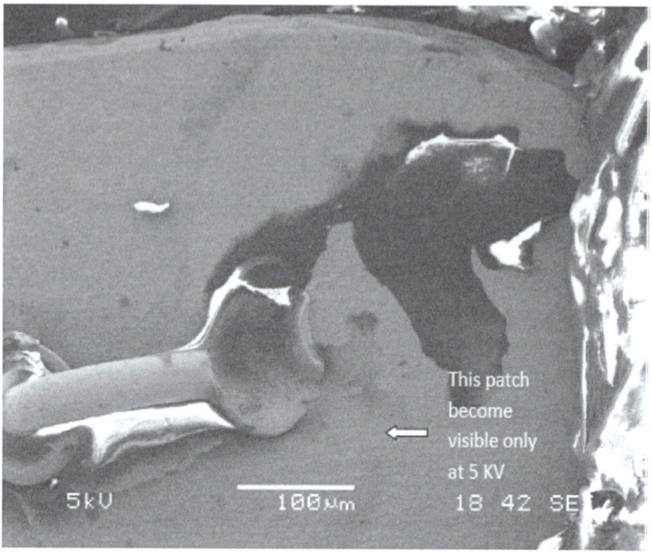


**Figure 6.10** SEM micrograph as Fig. 6.9, but with 5 KV accelerating voltage setting [2]. *(IEEE Transactions on Device and Materials Reliability, IEEE. Reprinted with permission.)*

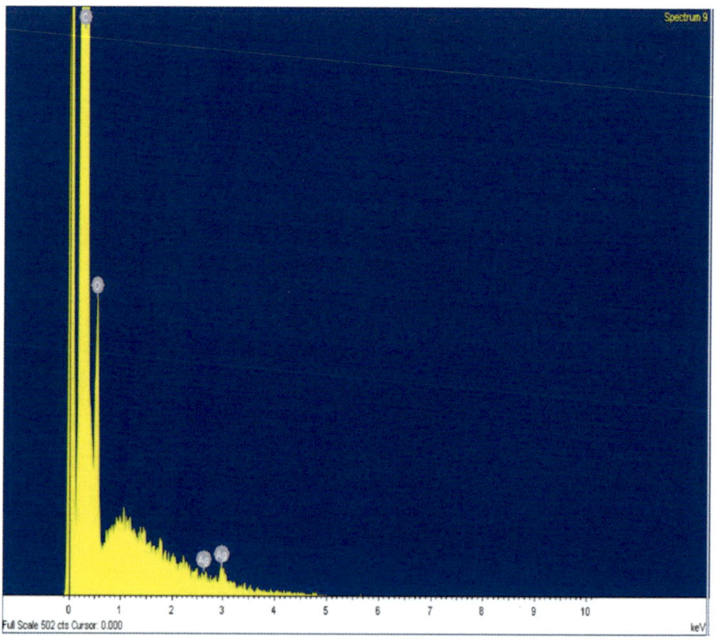


**Figure 6.11** At 5 KV, EDS spectrum on the contaminated layer (shown in Fig. 6.9) reveals very low Ag presence [2]. *(IEEE Transactions on Device and Materials Reliability, IEEE. Reprinted with permission.)*

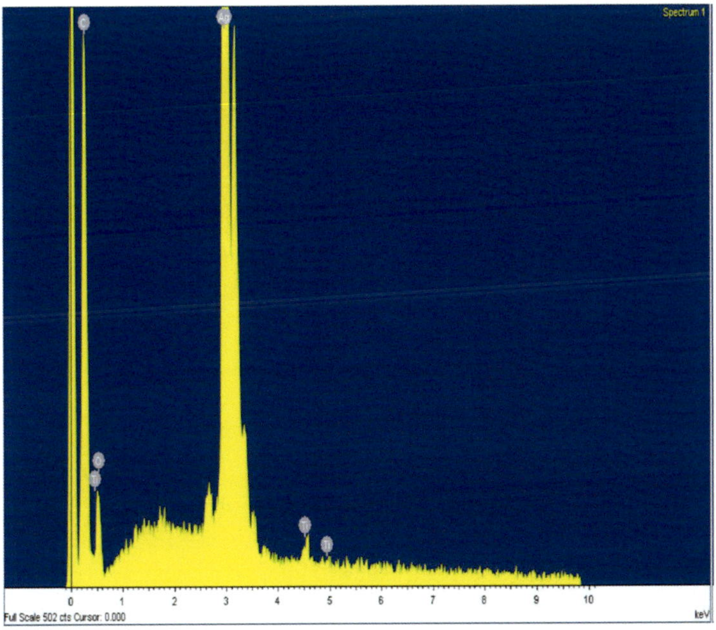


**Figure 6.12** At 15 KV, EDS spectrum on the contamination layer (shown in Fig. 6.9) reveals much higher Ag content [2]. *(IEEE Transactions on Device and Materials Reliability, IEEE. Reprinted with permission.)*

pad and dice temperature, changing the polymer composition and contaminating the Ag coating further. Due to the capillary tube effect, the incoming moisture at the pad region may also be vaporized as the temperature rises, causing it to flow along the dice–encapsulant interface. Chen et al. [18] found a similar phenomenon, which Tan et al. [9] verified using finite element analysis.

As a result, Tan et al. [2] hypothesized that as the moisture concentration decreases, the anodic decomposition of Ag will be slowed down, and water vapor will dissolve phosphor on the LED dice surface, causing phosphor degradation, and phosphor would fill the gaps at the interface between the bond wire and encapsulant or disperse into the encapsulant, alters the property of the encapsulant. The presence of the encapsulant's polymer material on the dice and the phosphor deteriorated layer, as seen in Figs. 6.13–6.16, supports the abovementioned postulation. Since decapping was difficult for these samples, good adhesion of dice and encapsulant was shown, and thus, separation of the encapsulant from the bond pad was easily seen, as shown in Figs. 6.9 and 6.10.

Delamination may become more extreme as more moisture penetrates the LED package under 95% RH, but due to the moisture saturation in the

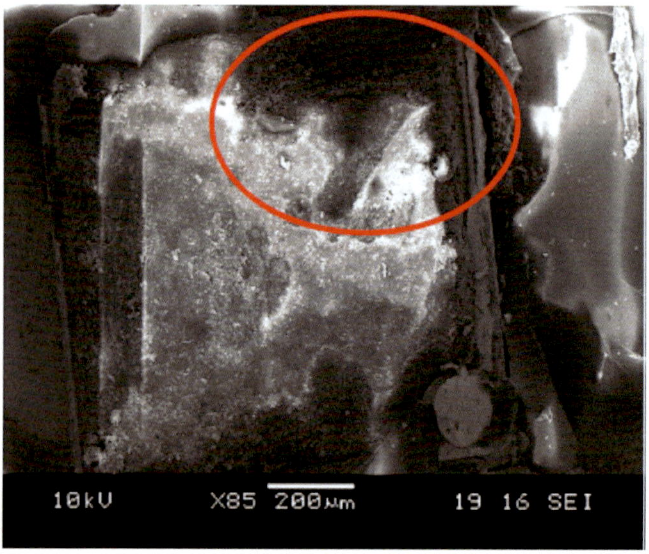


**Figure 6.13** SEM picture of decapped unit #10 after 560 h of 95% RH/85°C. The undissolved encapsulant that covered the bond pads is the large white spot [2]. *(IEEE Transactions on Device and Materials Reliability, IEEE. Reprinted with permission.)*

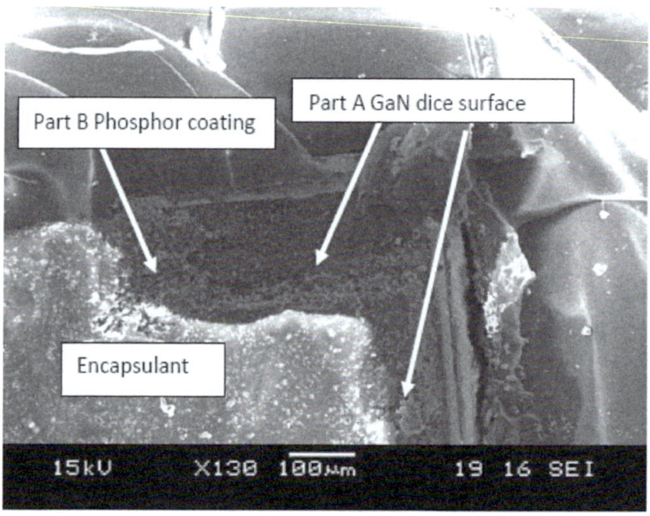


**Figure 6.14** Fig. 6.13, zone in of the red-circled area, indicating nonuniform phosphor breakdown on the dice surface. Some disintegration happens in the middle of the phosphor layer, indicating that vapor is traveling along the interface and reaching the center. Part A is the GaN dice, and Part B is the phosphor, as shown in the EDS spectrums below [2]. *(IEEE Transactions on Device and Materials Reliability, IEEE. Reprinted with permission.)*

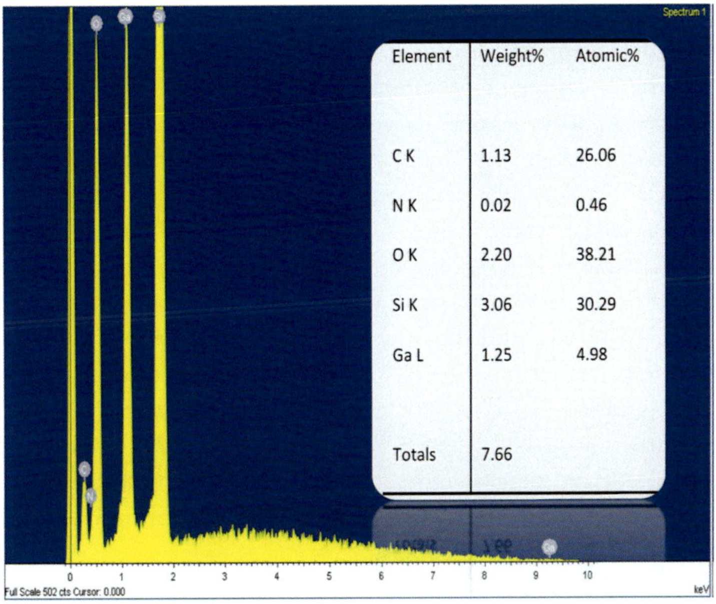


**Figure 6.15** Part A's EDS spectrum at 10 KV reveals that it is a GaN dice surface [2]. *(IEEE Transactions on Device and Materials Reliability, IEEE. Reprinted with permission.)*

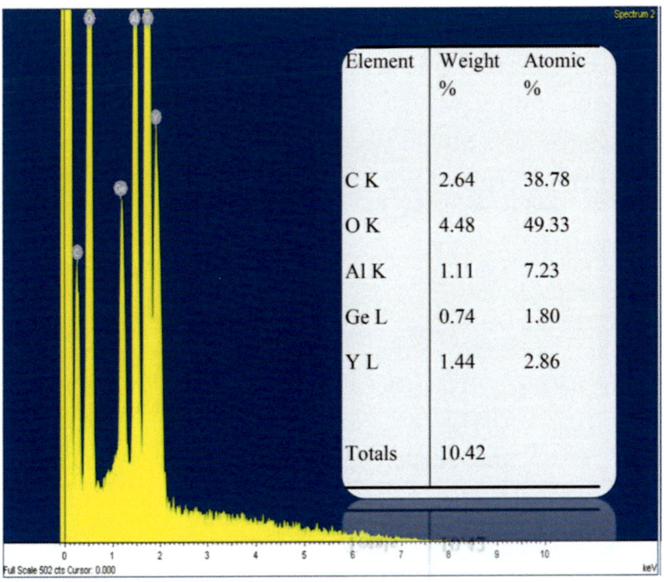


**Figure 6.16** EDS spectrum of Part B at 10 KV, showing that it is likely to be the phosphor coating.

encapsulant gel, it cannot diffuse any further. As a result, moisture will only fill the gap between the LED dice surface and the encapsulant, leading to more extreme encapsulant delamination from the LED dice surface.

The presence of moisture between the encapsulant and the LED dice caused more light dispersion and can even produce total internal reflection of the light from the dice due to the change in refractive index of the encapsulant and the water sheet. Both can cause more lumen degradation, as shown in Table 6.1, ranging from 10% to 15% under 95% RH.

As a result of hygroscopic stress produced by moisture absorption, Von Mises stress at the dice and die attachment has increased, as found by Tan et al. [9] in their earlier work, die attachment delamination may be another potential degradation mechanism from the 10% degradation stage. When samples were measured at 85°C/85% RH for 24 h, die attachment delamination and an increase in heat resistance from chip to heat sink were found by Hu et al. [19]. As a result, the LED junction temperature rose, resulting in lumen degradation.

Tan et al. [2] calculated the forward voltage ($V_f$) of the deteriorated LEDs at 300 mA during their research, and the results are shown in Table 6.8. Using a thermoelectric cooler, the ambient temperature was kept constant at 30°C for the measurements. Table 6.8 shows the measured $V_f$ values minus the voltage loss owing to Rs, as estimated previously by Tan et al. As a result, the values in Table 6.8 reflect the LED junction's actual $V_f$.

In the case of 95% RH (shown in shaded cells), $V_f$ decreases from 10% degradation level onward, which is possibly corresponding to the increase in junction temperature, which suggests the thermal resistance rise due to die attachment delamination (shown in shaded cells). However, the $V_f$ rise from 5% to 10%, and even 15%, for samples checked under 70% RH is yet to be investigated.

**Table 6.8** For the deteriorated units, forward voltage drop (Vf) of LEDs at 300 mA [2].

| Test conditions | Unit # | @ 5% degradation | @ 10% degradation | @ 15% degradation |
|---|---|---|---|---|
| 70%RH, 85 °C | 1 | 3.31 | 3.46 | 3.56 |
| | 18 | 3.34 | 3.47 | 3.54 |
| 95%RH, 85 °C | 6 | 3.49 | 3.69 | 3.61 |
| | 18 | 3.49 | 3.55 | 3.48 |

## 6.5 Summary

The time evolution of LED degradation physics under high temperature—humidity conditions was addressed in this chapter. Kirkendall voids, bond pad contamination, galvanic dissolving of Ag, phosphor dissolution, encapsulant delamination, and die attachment delamination are all considered as irreversible LED degradation processes. Before performing destructive analysis techniques, this chapter emphasizes the importance of statistical approaches and the use of nondestructive resources to recognize potential degradation mechanisms.

The obvious inference of this chapter is that the degradation mechanisms for high-power packed LEDs vary at various humidity levels, making extrapolation impossible, as in contrast to the case of integrated circuits. In order to estimate the LEDs' lifetime under humid operating conditions, new humidity models are needed. The research presented here used white high-power LEDs in an off state, and the function of the phosphor in converting blue light to white light, which generates heat, was not considered. The next chapter will compare the degradation mechanisms of blue and white packed LEDs under various test conditions.

## References

[1] P. Singh, C.M. Tan, Degradation physics of high power LEDs in outdoor environment and the role of phosphor in the degradation process, Scientific Reports 6 (1) (2016) 1—13.

[2] C.M. Tan, P. Singh, Time evolution degradation physics in high power white LEDs under high temperature-humidity conditions, IEEE Transactions on Device and Materials Reliability 14 (2) (2014) 742—750.

[3] E.A. Elsayed, Reliability Engineering, Addison-Wesley, Reading, MA, USA, 1996 ch. 6.

[4] P.A. Tobias, D.C. Trindade, Applied Reliability. New York, NY, USA: Van Nostrand, 1995 ch. 7.

[5] F.J. Yeh, T.-C. Chiu, K.-L. Lin, The interfacial interaction of Ti/Ni/Ag/Au Multilayer under thermal cycling test, in: Proc. 14th Int. Conf. EMAP, 2012, pp. 1—5.

[6] C.M. Tan, Z. Gan, W.F. Ho, S. Chen, R. Liu, Determination of the dice forward *I—V* characteristics of a power diode from a packaged device and its applications, Microelectronics Reliability 45 (1) (January 2005) 179—184.

[7] General Information on the Assembly and Solder Pad Design of the DRAGON Family, Application Note from OSRAM. [Online]. Available: http://catalog.osram-os.com/catalogue/catalogue.do;jsessionid=4B38A203E32C13293C7E445DFE6B4F55?favOid=000000010002a46a043f0023& act=showBookmark.

[8] W.C. Mallard, A.B. Gardner, R.F. Bass, L.M. Slifkin, Selfdiffusion in silver-gold solid solutions, Physical Review 129 (2) (January 1963) 617—625.

[9] K. Sugawara, M. Kawamura, Y. Abe, K. Sasaki, Comparison of the agglomeration behaviour of Ag(Al) films and Ag(Au) films, Microelectronic Engineering 84 (11) (November 2007) 2476—2480.

[10] X. Luo, B. Wu, S. Liu, Effects of moist environments on LED module reliability, IEEE Transactions on Device and Materials Reliability 10 (2) (June 2010) 182–186.
[11] L.X. Tan, J. Li, K. Wang, S. Liu, Effects of defects on the thermal and optical performance of high-brightness light-emitting diodes, IEEE Transactions on Electronics Packaging Manufacturing 32 (4) (October 2009) 233–240.
[12] P.R. Roberge, Handbook of Corrosion Engineering, McGraw-Hill, New York, NY, USA, 2012 ch. 5.
[13] V. Brusic, G.S. Frankel, J. Roldan, R. Saraf, Corrosion and protection of a conductive silver paste, Journal of the Electrochemical Society 142 (8) (August 1995) 2591–2594.
[14] K. Vu, Silver Migration—The Mechanism and Effects on Thick Film Conductors. Available Online: http://www.researchgate.net/publication/242074155_Silver_Migration_The_Mechanism_and_Effects_on_Thick-Film_Conductors.
[15] S. Mena, Electrochemical Migration of Silver under Temperature Humudity- Bias-Stressing-Conditions, M.S. thesis, Univ. Philippines, Quezon City, Philippines, 2010.
[16] D.L. Barton, M. Osinski, P. Perlin, P.G. Eliseev, J. Lee, Singlequantum well InGaN green light emitting diode degradation under high electrical stress, Microelectronics Reliability 39 (8) (August 1999) 1219–1227.
[17] G. Meneghesso, et al., Reliability of visible GaN LEDs in plastic package, Microelectronics Reliability 43 (9–11) (September–November 2003) 1737–1742.
[18] S.H. Chen, C.M. Tan, G.H. Tan, F.F. He, Degradation behavior of high power light emitting diode under high frequency switching, Microelectronics Reliability 52 (9/10) (September/October 2012) 2168–2173.
[19] J.Z. Hu, L. Yang, M.W. Shin, Mechanism and thermal effect of delamination in light-emitting diode packages, Microelectronics Journal 38 (2) (February 2007) 157–163.

# CHAPTER 7

# Comparison of blue and white packaged LED degradation

## 7.1 Introduction

The unique time-varying lumen degradation patterns for high-power LEDs (HP- LEDs) under extreme environmental conditions with LEDs in the OFF condition were discussed in Chapters 4—6. LEDs face extreme temperatures, humidity, and electrical stresses concurrently in many real-world applications, but reliability tests that include all these three stresses are uncommon, as discussed in the previous chapters.

The impact of all these stresses acting simultaneously on the LED's reliability will be studied in this chapter. Singh and Tan devised a test for this purpose, dubbed the MET (Moisture-Electrical-Thermal test) [1], to investigate the physics of degradation. They compared the LEDs failure modes in the MET test to those in the literatures for conventional single or double stresses. The dependency of phosphor on the MET test of HP-LEDs was also compared to the degradation of blue LEDs (i.e., LEDs without phosphor layer).

## 7.2 Experimentation

In the work of Singh and Tan, they selected two sets of OSRAM golden dragon LEDs with 40 blue and 40 white LEDs each. As shown in Table 7.1, the two sets were further divided into two subsets, each with 20 LEDs [1]. According to the manufacturing specification, LEDs are powered on with 350 mA constant current. White and blue LEDs with power-off conditions were used in subsets B1 and B2. For an accelerated life test, all sets were put in a temperature-humidity chamber (series from Isuzu) with a test temperature of 85°C and an RH of 85%. This state is intended to resemble real-world outdoor conditions more closely in humid countries, as well as marine and aerospace settings [1].

*Reliability and Failure Analysis of High-Power LED Packaging*
ISBN 978-0-12-822408-3
https://doi.org/10.1016/B978-0-12-822408-3.00006-X

**Table 7.1** Test conditions of LEDs for experimentation.

| | White LEDs | | Blue LEDs | |
|---|---|---|---|---|
| **Test condition** | **A1** | **B1** | **A2** | **B2** |
| Power status | ON (350 mA) | OFF | ON (350 mA) | OFF |
| Environmental conditions | 85°C and 85%RH | 85°C and 85%RH | 85°C and 85%RH | 85°C and 85%RH |

P. Singh, C.M. Tan, Degradation physics of high power LEDs in outdoor environment and the role of phosphor in the degradation process, Scientific Reports 6 (1) (2016), 1–13. 

Keithley source meter model 2651A was used to take electrical measurements on all the LEDs. Their optical measurements were made with a setup that included a 1-m-diameter integrating sphere model SLM-40TS-110902 and an Ocean Optics QE6500 spectroradiometer. All of them had an initial collection of measurements taken to act as a reference benchmark for all samples.

For optical and electrical tests, the samples were removed from the humidity-temperature chamber after every 24 h. To avoid self-heating during measurements, all measurements were set according to the work of Tan et al. [2]. The electrical characterization of these LEDs was also compared under various conditions.

## 7.3 Statistical analysis on the test results

Fig. 7.1 depicts the lumen degradation with respect to time for blue LEDs evaluated under both different situations, as well as white LEDs tested under solely ON settings. During the MET test, white LEDs with the power turned off showed no noticeable deterioration.

As shown in Fig. 7.1, the entire test duration can be observed as three stages. The first stage is the early stages of degradation, when moisture entering the silicone encapsulation was identified as the primary cause of lumen degradation. After the initial degradation, the next stage represents the recovery period, where a rapid degradation fall is observed. Due to the extensive monitoring, the last stage indicates the final permanent deterioration stage. Table 7.2 lists the durations of each of the three stages for both colored LEDs under various environmental conditions.

In the initial degradation stage, blue LEDs when stressed under ON conditions degraded in the first 24 h and up to 12%, while they degraded

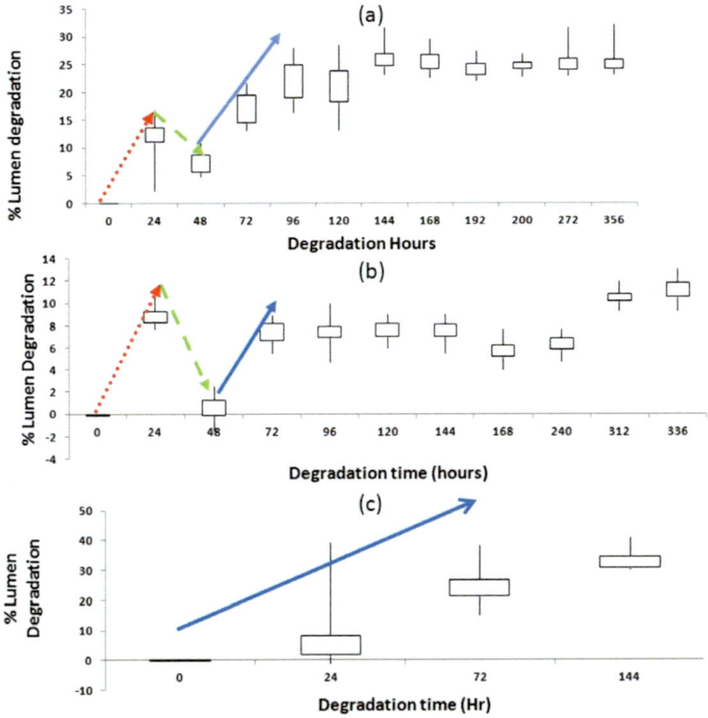


**Figure 7.1** Percentage lumen degradation with respect to test duration for LEDs measured under varied operation situations. (A) Blue LEDs under ON state; (B) Blue LEDs under OFF state; and (C) White LEDs under ON state. For LED, the red line represents initial degradation, the green represents lumen recovery, and the line represents final degradation. *(P. Singh, C.M. Tan, Degradation physics of high power LEDs in outdoor environment and the role of phosphor in the degradation process, Scientific Reports 6 (1) (2016), 1–13.* © Scientific Reports, *Nature. Reprinted with permission.)*

**Table 7.2** The time taken for different stages by the blue and white LEDs under various working conditions.

| Samples/ Stages | Blue LED ON | Blue LED OFF | White LED ON | White LED OFF |
|---|---|---|---|---|
| Initial degradation | 12% lumen degradation | 8.5% lumen degradation | Nil | No degradation |
| Recovery stage | 7.5% degradation | 1% lumen degradation | Nil | No degradation |
| Final degradation | 30% lumen degradation | 12% lumen degradation | 33% lumen degradation | No degradation |

P. Singh, C.M. Tan, Degradation physics of high power LEDs in outdoor environment and the role of phosphor in the degradation process, Scientific Reports 6 (1) (2016), 1–13. © *Scientific Reports*, Nature. Reprinted with permission.

just 8.5% (in the first 24 h) when they were in the OFF state, as shown in Table 7.2. Under both ON and OFF conditions, blue LEDs showed a recovery trend in its lumen degradation within the initial 24 h. In the next 24 h of testing, blue LEDs showed a recovery trend of lumen degradation falling down to 7.5% under ON state, and it falls to 1% under OFF state. LEDs began to degrade again after the recovery stage. At this stage, 30% of lumen degradation was experienced by the blue LEDs after 356 h under ON state, while 12% lumen degradation under off state after 336 h [1].

Under the ON condition, white LEDs did not display any signs of recovery, and it reaches 33% lumen degradation in 144 h, and the test was stopped then, in accordance with the ASISST standard [3], which set the maximum permissible LEDs degradation at 30%. This shows that phosphor has a negative impact in white LEDs lifetime as compared to blue LEDs in an outdoor setting as it is the only difference between two types of LEDs tested in this work. Phosphor can generate extra heat during the light conversion process as shown by Refs. [4,5], and higher amount of heat in the white LEDs due to phosphor was determined to be the culprit for rapid lumen degradation in white LEDs when compared to blue. However, in the recent years, several attempts have been made to reduce this negative effect of heat generation by phosphor in LEDs.

White LEDs, however, experienced almost no lumen degradation under power OFF conditions, whereas blue LEDs experienced 12% degradation after 336 h, as presented in Table 7.2. This disparity in deterioration was due to white LEDs' shorter test period, and the reason behind the negligible decrease in white LED OFF condition remains unknown, Fig. 7.2 shows that a finite element (FEM) study of moisture penetration into LED package, as described earlier in Chapter 5, can provide a much clearer explanation.

At the start, moisture was being absorbed into the encapsulant, as shown in Fig. 7.2. At about 132 h, moisture gets saturated inside the LEDs encapsulant and thereafter moisture starts penetrating inside the die attach. Before 132 h, very less moisture reached die attach. As more moisture starts reaching die attachment of the LEDs, delamination occurs, which results in LEDs thermal resistance (Rth) and thus LEDs lumen decreased, as shown by Ref. [6]. As shown in Fig. 7.2, die-attachment delamination started at about 150 h. As heat generated by phosphor in white LEDs can evaporate incoming moisture partially, the die attachment showed better strength in white LEDs and less light attenuation observed for white light LEDs. As a result, there was no major lumen degradation.

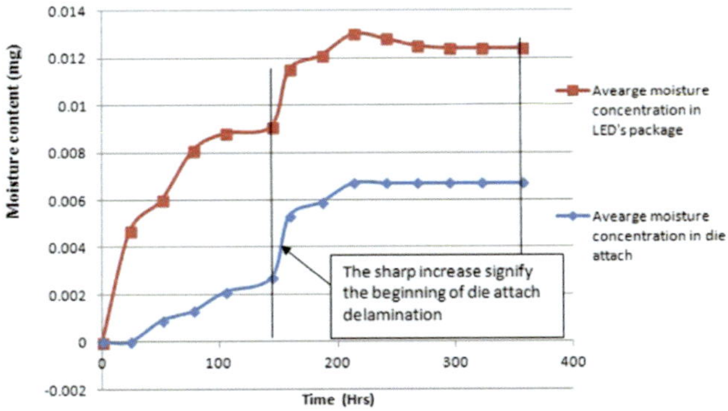


**Figure 7.2** Finite element analysis (FEM) revealed the moisture penetration amount into LED packages and die attach. *(P. Singh, C.M. Tan, Degradation physics of high power LEDs in outdoor environment and the role of phosphor in the degradation process, Scientific Reports 6 (1) (2016), 1–13.* © Scientific Reports, *Nature. Reprinted with permission.)*

Die attachment delamination had already occurred in blue LEDs examined for 336 h, and the moisture contained inside the die attachment cannot be evaporated during further operation, resulting in substantial degradation as observed. The blue LEDs moisture content was 40% higher than white LEDs, with the difference being entirely due to the die attachment content. As can be seen in Table 7.2, LEDs under power-on conditions degraded faster than those under power-off conditions, as predicted. To better understand the mechanism of the increase in the percentage of decay of the lumen shown in Fig. 7.1, and to examine the effect of phosphors present in the white LED, the structure of the LEDs under study needs to be clarified, and it is shown in Fig. 7.3.

LEDs have a molding component for mechanical integrity (shown as molded housing in Fig. 7.3) that also contains the LED chip, encapsulation, lead frame or bonding wire, and heat sink (for heat dissipation). Since heat sink is inserted within the molding portion underneath the LED die, and it is connected with the LED die via die attachment, it is not visible in Fig. 7.3. The molding part is made of hard silicone, while the encapsulation comprises silicone in rubber form [7]. This molding component works as reflector source, enhancing overall LEDs lumen production while protecting the internal LED components from mechanical and environmental stress. The silicone encapsulation directs the light beams while also protecting the LED internal components from external stresses [1].

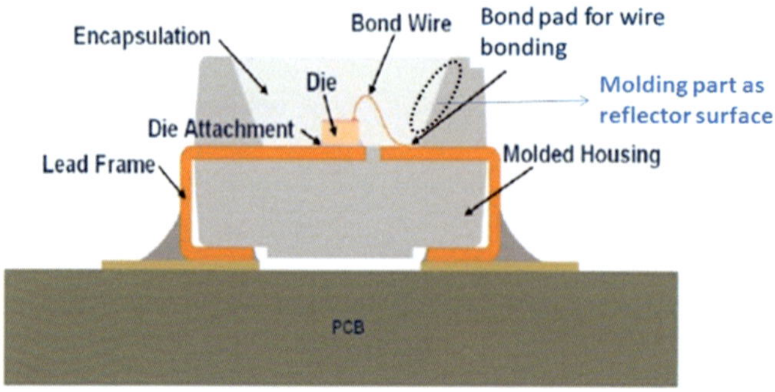


**Figure 7.3** Materials and structure used in LEDs under study. *(P. Singh, C.M. Tan, Degradation physics of high power LEDs in outdoor environment and the role of phosphor in the degradation process, Scientific Reports 6 (1) (2016), 1—13.* © Scientific Reports, *Nature. Reprinted with permission.)*

Poly-phenylsiloxanes and poly-dimethylsiloxane are combined in various amounts to transform silicone into different types, according to the OSRAM datasheet. The molding component and the encapsulation silicone attain different coefficients of thermal expansion due to the difference in material concentrations [8,9]. Fig. 7.4 shows the thermal expansivities of the two parts determined with a dilatometer with model NETZSCH DIL 402 PC.

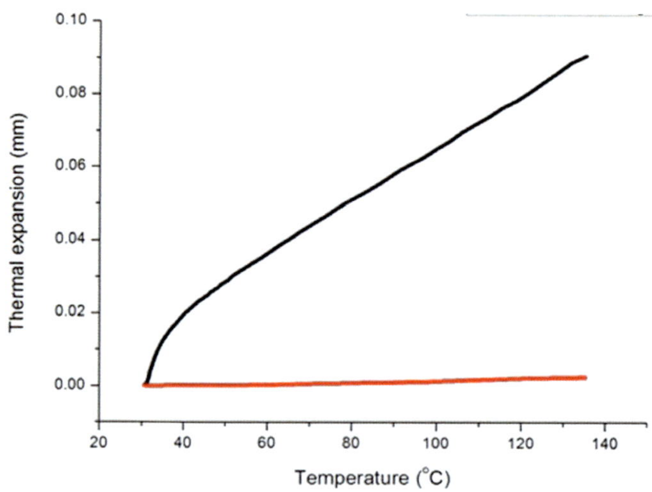


**Figure 7.4** Thermal expansion coefficient of molding and sealant measured at different temperatures. *(P. Singh, C.M. Tan, Degradation physics of high power LEDs in outdoor environment and the role of phosphor in the degradation process, Scientific Reports 6 (1) (2016), 1—13.* © Scientific Reports, *Nature. Reprinted with permission.)*

## 7.4 Physical analysis on the test samples

The chamber temperature was 85°C during the MET test, but the internal temperature of LEDs can exceed 135°C when they are switched on, according to the OSRAM datasheet [10]. LED's molding part and encapsulant expanded at this high temperature during testing, with encapsulant experiencing 97.5% more expansion than molding part at such high temperatures [10].

The thermal expansivity difference of the molding component and the encapsulant causes differential thermal and mechanical stress at the interface of two components, resulting in generation of voids/cracks at the interface between the two parts. Cracks were apparent under a high-resolution optical microscope for blue and white LEDs, respectively, as shown in Fig. 7.5A and B [1]. These cracks and delamination allow moisture to easily diffuse into LEDs, resulting in degradation of light output.

Singh et al. [1] chose a sample from each collection experiencing the highest forward voltage (Vf) decrease for C-SAM analysis to verify that the delamination occurred was indeed the cause behind light output degradation. Fig. 7.6 shows the standard IV curves of the two types of LEDs that are observed under different tests. Under constant current, the decrease in the forward voltage of the LED means that the thermal resistance of the LED chip has deteriorated due to the stratification of the die attach [11,12], thus the maximum decrease in the forward voltage is selected as the selection criterion for the tested LEDs for C-SAM examination.

Another two blue LEDs samples tested from the ON and OFF tests are also selected to verify the effect of the MET test before the recovery process since recovery process of light output degradation was seen during the MET tests [1]. The same samples were used to screen for potential delamination during the C-SAM examination. Table 7.3 shows the test conditions that these six samples were subjected to, as well as the reasoning behind their decisions.

Under C-SAM, square in the LED package center in Fig. 7.7 is the LED die attachment region.

Delamination occurred at the silicone encapsulant and the molding component interface was clearly visible in C-SAM images of white LEDs tested under ON conditions, as seen through the red marked areas in Fig. 7.7B. As samples were exposed to high temperatures, the red areas indicated where cracks/delamination occurred. This demonstrated that delamination was the reason for lumen degradation in white LEDs tested under ON conditions, and that this delamination was caused by internal heat generation.

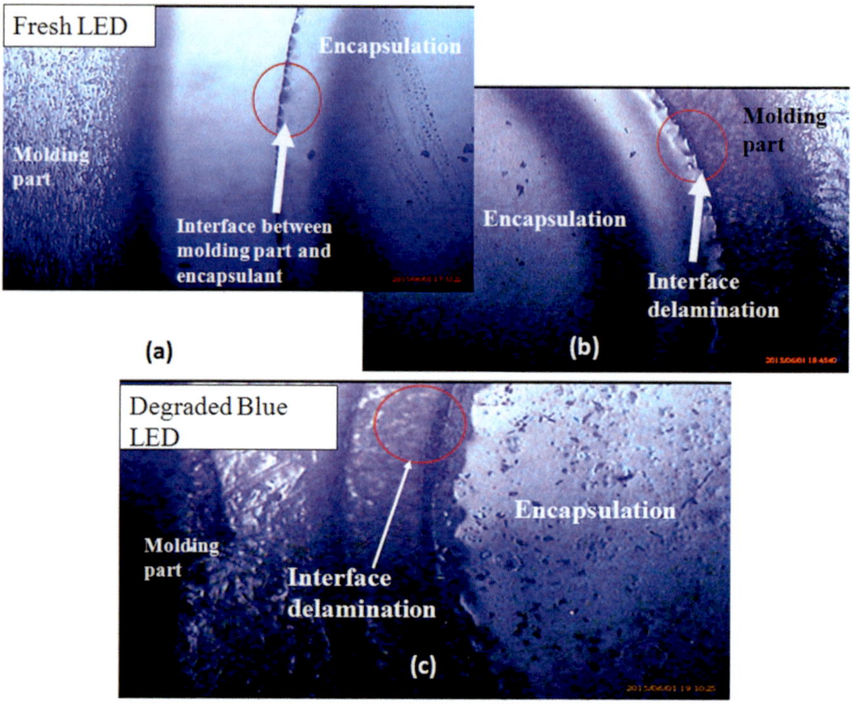


**Figure 7.5** Light micrographs show that compared to (A) newer LEDs, (B) white LED and (C) blue LED have delamination due to the difference in thermal expansion at 5× magnification. The optical microscope used by is Keyence 5000. *(P. Singh, C.M. Tan, Degradation physics of high power LEDs in outdoor environment and the role of phosphor in the degradation process, Scientific Reports 6 (1) (2016), 1–13.* © Scientific Reports, *Nature. Reprinted with permission.)*

The delamination of blue LEDs under ON conditions was similar to the delamination of white LEDs under ON conditions, as shown in Fig. 7.8. In comparison of Figs. 7.7 and 7.8, the blue LEDs had more delamination than that of the white LEDs because the red region was larger. This is likely due to the fact that the blue LEDs were tested for 356 h, while white LEDs were only tested for 144 h. The state of blue LEDs under OFF conditions, on the other hand, showed no noticeable delamination, suggesting that the delamination was caused by the chip's internal heat.

## 7.5 Finite element analysis

Singh and Tan [1] used FEM to obtain a quantitative knowledge of their proposed mechanism. The finite element analysis was performed using

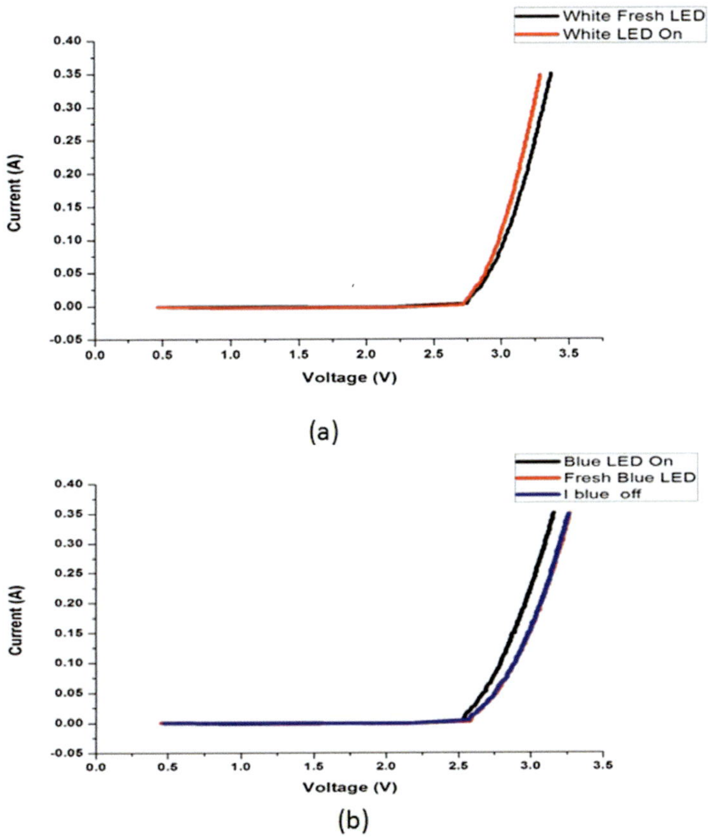


**Figure 7.6** IV characteristics of (A) white and (B) blue LEDs measured after testing in different environmental conditions. *(P. Singh, C.M. Tan, Degradation physics of high power LEDs in outdoor environment and the role of phosphor in the degradation process, Scientific Reports 6 (1) (2016), 1–13.* © Scientific Reports, *Nature. Reprinted with permission.)*

ANSYS. Fig. 7.9 depicts the LED structure used in the ANSYS simulation. For ease of verification, only one-fourth of the entire LEDs structure was used.

The temperature provided to the external body was 85°C and at the LED chip surface, it was 135°C, as stated previously, for the simulation when LEDs were tested under ON condition. Table 7.4 describes the material parameters values used in FEM [13].

Fig. 7.10 shows the mechanical strains and von Mises stresses distribution in the LED under the two previously discussed conditions.

**Table 7.3** Conditions and reasons for choosing six samples for venous examination and CSAM spectroscopy.

| Unit # | Test condition undergone by the test unit | White/ Blue LEDs | Rationale of the choice | Remarks |
|---|---|---|---|---|
| 1 | Power on-condition, tested for 144 h | White | Maximum degraded sample with 40% lumen depreciation | Test stopped after 144 h as LEDs under on condition reached 70% of its lumen output, following the ASSIST standards |
| 2 | Power off-condition, tested for 144 h | White | Maximum degraded sample with 2% lumen depreciation | |
| 3 | Power on-condition, tested for 356 h | Blue | Maximum degraded sample with 32% lumen depreciation | Test stopped after 356 h as LEDs under on condition reached 70% of its lumen output, following the ASSIST standards |
| 4 | Power off-condition, tested for 356 h | Blue | Maximum degraded sample with 13% lumen depreciation | |
| 5 | Power on-condition, tested for 24 h | Blue | Maximum degraded sample with 13% lumen depreciation | Samples are chosen from the respective sets to observe the effect of MET test before and after degradation recovery |

P. Singh, C.M. Tan, Degradation physics of high power LEDs in outdoor environment and the role of phosphor in the degradation process, Scientific Reports 6 (1) (2016), 1–13. 

The maximum mechanical stress appears at the LED chip–encapsulant interface when the LED is on, and it appears on the outer surface of the encapsulant when the LED is off, as shown in Fig. 7.10. Under both power ON and power OFF conditions, the mechanical strain at the encapsulant–molding interface was very high. The maximum value of mechanical strain

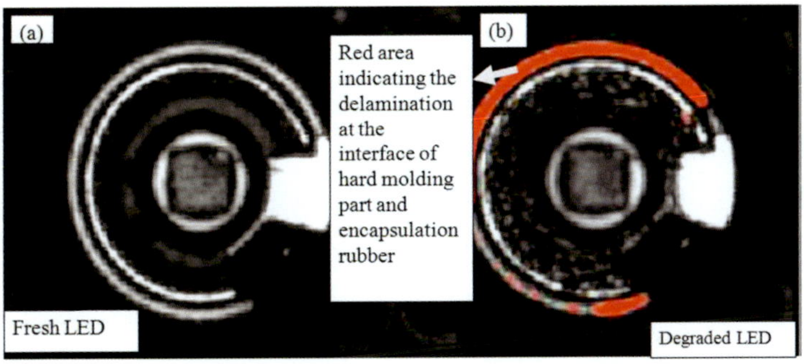


**Figure 7.7** The C-SAM image shows delamination at the white LED silicone and the molded part interface. Red area indicates the location of the cracks when the sample is exposed to high temperatures. The image on the left shows the C-SAM image of the new LED, and the image on the right is the C-SAM image of the degraded LED (Example 1 in Table 7.3). The chip area is identified by the square in the LED package center. *(P. Singh, C.M. Tan, Degradation physics of high power LEDs in outdoor environment and the role of phosphor in the degradation process, Scientific Reports 6 (1) (2016), 1–13.* © Scientific Reports, *Nature. Reprinted with permission.)*

and von Mises stresses at chip/encapsulant and encapsulant/molding interface for LEDs tested in ON test is substantially higher than LEDs tested in OFF test, as summarized in Table 7.5. Furthermore, when the LED was turned off, the highest von misses stress was found at the copper heat sink's outer edge, which was far from the abovementioned interfaces. Consequently, ON state LEDs were more likely to face delamination in encapsulant and chip interface, which resulted in an opening that allowed moisture particles to easily penetrate into the package. This result was consistent to the observations under optical and C-SAM inspections.

## 7.6 Uniqueness of degradation mechanisms for blue and white LEDs

The electrical test results as summarized in Fig. 7.6 were also consistent with the above findings [1]. The samples were subjected to I–V measurements for comparison between the LEDs electrical characteristics examined when tested in various conditions. Forward voltages decreased for all tested samples, as shown in Fig. 7.6. This can be demonstrated in the following way.

Moisture entered the LEDs die attachment material quickly when they were tested under ON condition due to the molding–encapsulation interface delamination, which led to LED die and die attachment

**Figure 7.8** The C-SAM image shows delamination at the interface between the gradient BLUE LED silicone and the molded part. The red marked area shows the crack/delamination occurrence when the sample is tested under high temperatures. The image on the left shows the C-SAM image of the new LED, and the image on the right is the C-SAM image of the degraded LED (Example 3 in Table 7.3). The chip area is observed by the square in the LED package center. *(P. Singh, C.M. Tan, Degradation physics of high power LEDs in outdoor environment and the role of phosphor in the degradation process, Scientific Reports 6 (1) (2016), 1–13.* © Scientific Reports, *Nature. Reprinted with permission.)*

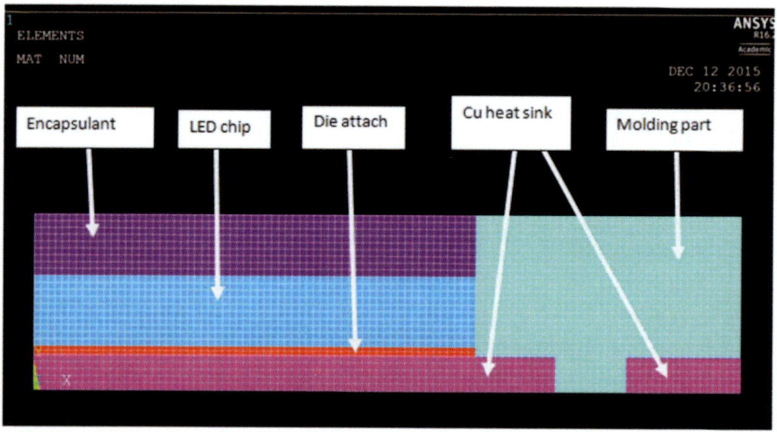


**Figure 7.9** LED design employed for Ansys simulation. *(P. Singh, C.M. Tan, Degradation physics of high power LEDs in outdoor environment and the role of phosphor in the degradation process, Scientific Reports 6 (1) (2016), 1–13.* © Scientific Reports, *Nature. Reprinted with permission.)*

delamination with a resulting forward voltage decrease, as observed for all the LEDs degraded in power ON state.

**Table 7.4** Material parameters values used in FEM.

| Material | Density ($kg/m^3$) | Specific heat (J/kg*K) | Thermal conductivity (W/m*K) | Elastic modulus (GPa) | Poisson ratio | Coefficient of thermal expansion ($e^{-6}$/°C) |
|---|---|---|---|---|---|---|
| Molding part | 1100 | 1591 | 0.209 | 2.3 | 0.394 | 80 |
| Encapsulant | 1200 | 1260 | 0.2 | 1 | 0.4 | 220 |
| Die attach | 7390 | 217 | 57 | 52.6 | 0.35 | 20 |
| LED chip | 3965 | 730 | 130 | 210 | 0.17 | 7.75 |
| Copper heat sink | 8950 | 390 | 400 | 110 | 0.34 | 18 |

P. Singh, C.M. Tan, Degradation physics of high power LEDs in outdoor environment and the role of phosphor in the degradation process, Scientific Reports 6 (1) (2016), 1–13. © *Scientific Reports*, Nature. Reprinted with permission.

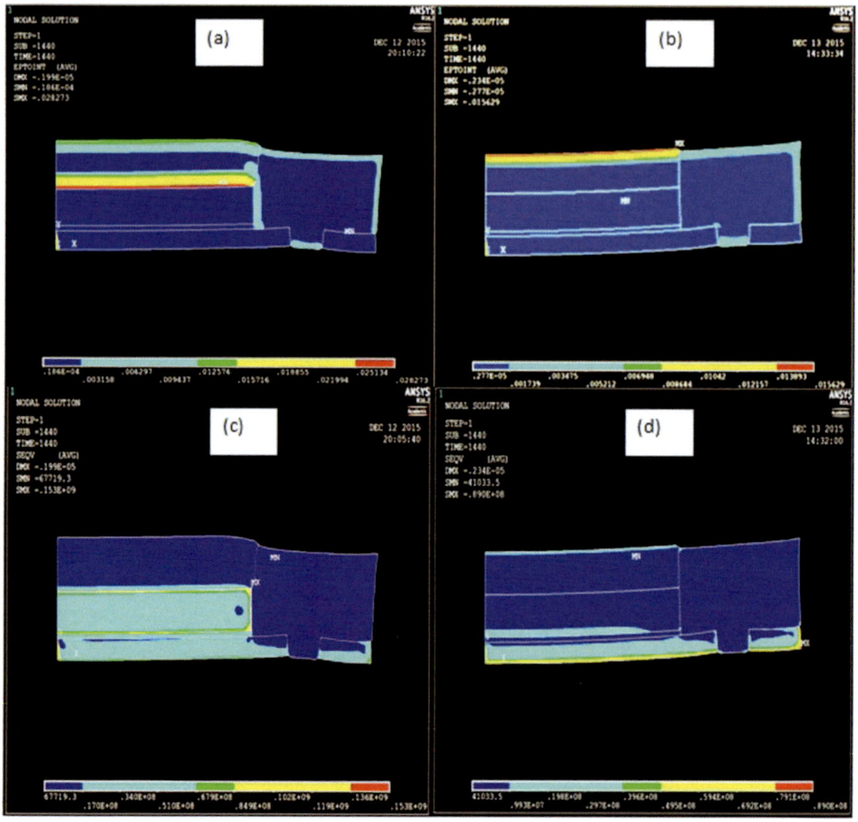


**Figure 7.10** Correlation of mechanical strain in (A) LED under ON state and (B) LED under OFF state, and von misses pressure in (C) LED under ON state and (D) LED under OFF state. *(P. Singh, C.M. Tan, Degradation physics of high power LEDs in outdoor environment and the role of phosphor in the degradation process, Scientific Reports 6 (1) (2016), 1–13.* © Scientific Reports, *Nature. Reprinted with permission.)*

**Table 7.5** Comparison between mechanical strain and von-misses stress at the chip and the package molding interface under LED on and off conditions.

| | LED chip- encapsulant interface | | Encapsulant- molding part interface | |
|---|---|---|---|---|
| | **LED ON state** | **LED OFF state** | **LED ON state** | **LED OFF state** |
| Max mechanical strain | 0.02 | 0.27E-5 | 0.009 | 0.27E-5 |
| Max von-misses stress (Pa) | 0.15E9 | 4.1E4 | 6.7E4 | 4.1E4 |

P. Singh, C.M. Tan, Degradation physics of high power LEDs in outdoor environment and the role of phosphor in the degradation process, Scientific Reports 6 (1) (2016), 1–13. © *Scientific Reports*, Nature. Reprinted with permission.

Ideality factor, series resistance, and reverse saturation current (n, Is, and Rs, respectively) of the LEDs were measured using an integration-dependent parameters extraction method established by Tan et al. [1,3,13] to verify the impact of harsh test conditions on the electrical characteristics of LED in this MET test. These variables offered valuable information about the packed high-power LEDs that were checked under various conditions. Tables 7.6 and 7.7 show the parameters extraction results.

Tables 7.6 and 7.7 show that the white light LED n, Is, and Rs values have not changed by an order of magnitude under the different test conditions, which shows that no LED chip faults exists in both cases. Thus the degradation mechanism is as follows.

Since the phosphor was additionally a heat generation source in white LEDs [14–16], moisture penetrating inside the package through the encapsulation/molding interface delamination will be pushed away due to the internal warmth due to the phosphor layer, which leaves behind negligible moisture on the surface of chip; therefore, Is values after the test remain unaltered. In the MET test, the phosphor layer that converts blue into yellow light in white light producing heat, which seems to protect the LEDs from moisture-related failures. However, the white LED's degraded more rapidly than the blue LEDs due to higher internal temperature generated due to both the chip and phosphor.

The blue LEDs, on the other hand, were not affected in this way. The value of $I_s$ decreased significantly when the power was turned on, but increased significantly when the power was turned off. For blue LEDs tested in ON state for 365 h, $I_s$ was reduced by 1000 times, while very slight

**Table 7.6** n, Rs, and Is values extracted from I–V curves for the tested packaged white LEDs.

| | Fresh white LED | White LED under power-off condition after 144 h | White LED under power-on condition after 144 h |
|---|---|---|---|
| n | 2.39 | 2.35 | 2.32 |
| Is (A) | 2.35E-22 | 1.7E-22 | 2.08E-22 |
| Rs (ohm) | 1.16 | 0.97 | 0.84 |

P. Singh, C.M. Tan, Degradation physics of high power LEDs in outdoor environment and the role of phosphor in the degradation process, Scientific Reports 6 (1) (2016), 1–13. © *Scientific Reports*, Nature. Reprinted with permission.

**Table 7.7** n, Rs and Is value extracted from I–V curves for the tested packaged blue LEDs.

| | Blue LED fresh | Blue LED under power-on condition after 356 h | Blue LED under power-off condition after 356 h | Blue LED under power-on condition before recovery (tested for only 24 h) | Blue LED under power-off condition before recovery (tested for only 24 h) |
|---|---|---|---|---|---|
| n | 2.44 | 2.09 | 3.83 | 2.33 | 2.81 |
| Is (A) | 1.83E-20 | 7.43E-23 | 7.95E-14 | 5.95E-23 | 5.03E-19 |
| Rs (ohm) | 0.96 | 1 | 0.87 | 1.13 | 0.92 |



change occurred for n and Rs. Singh and Tan [1] speculated that this was a result of ionic contamination at new LEDs junction, which was removed by the moisture that entered the junction. High junction temperature (Tj) of LEDs tested under ON condition, on the other hand, evaporated moisture present at the LED junction away toward the environment, which in turn eliminated contamination, which was dissolved in moisture, lowering reverse leakage current of LEDs. Fig. 7.8 shows the moisture presence at the perimeter of the blue LED chips when they were turned off during test, but it is not the case for the LED chips when they were turned on during test. When blue LEDs were turned off, moisture accumulated around the perimeter of the LEDs and remained due to insufficient heat for moisture evaporation. Thus, under power ON conditions, the LED chip was shielded from incoming moisture, but the heat too affected the die attachment and surrounding packaging materials, resulting in higher degradation when compared to blue LEDs under power OFF state. The increase in the thermal resistance of LEDs tested under the power ON was clearly indicated by the decrease in Vf as in Fig. 7.6(b).

### 7.6.1 Investigation of lumen recovery stage in LEDs

LEDs tested within 24 h, i.e., only before the recovery period, and also after the recovery duration (verified until the end) were examined with C-SAM to check the moisture penetration into blue LED packages. On comparing with LEDs after end of the test in Fig. 7.11B, very little

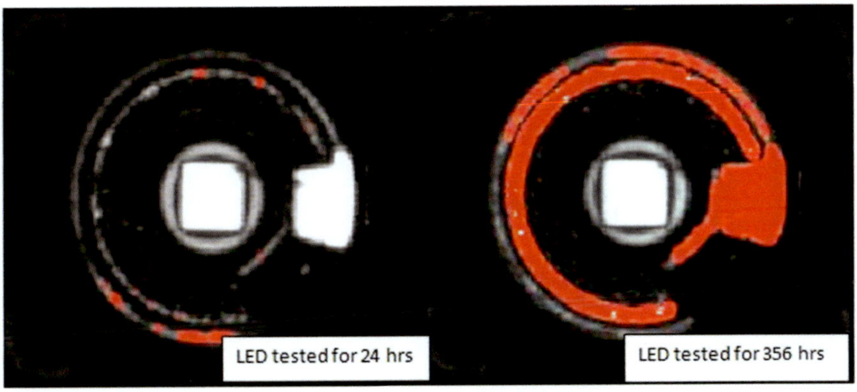


**Figure 7.11** C-SAM micrographs demonstrating blue LEDs condition before and after the recovery phase. Red color shows that cracking/delamination occurs. Left-side image shows a C-SAM image of a blue LED tested for 24 h, and the right-side image is a C-SAM image of a blue LED tested for 356 h. *(P. Singh, C.M. Tan, Degradation physics of high power LEDs in outdoor environment and the role of phosphor in the degradation process, Scientific Reports 6 (1) (2016), 1–13.* © Scientific Reports, *Nature. Reprinted with permission.)*

delamination was noticeable in the initial stage for LEDs, as seen in Fig. 7.11A. According to Tan et al. [4], the initial degradation was caused by moisture trapped in the silicone encapsulation, while the later percentage lumen degradation was caused by a large moisture amount entering through the gap formed at the silicone encapsulation–molding interface, resulting in various stages of degradation [1].

C-SAM examination for LEDs that were flipped upside down (i.e., lens down) was recorded for blue LEDs when they were before recovery. Small gaps or trapped bubbles were observed in the C-SAM images at the interface between the die mold/fixture and the LED die/chip interface (Fig. 7.12).

### 7.6.2 Comparison of LEDs' performance under MET test and other tests with single or multiple stress factors

There have been several reliability tests on high-power LEDs, but the majority of them used either a single or double stress effects on the LEDs. Individual stresses are typically electrical or operating temperature, while multiple stresses are a mixture of electrical with high temperatures or high temperature with humidity measurements. Table 7.8 summarizes the mechanisms and modes associated with the failure due to these multiple stresses along with individual stress on LEDs.

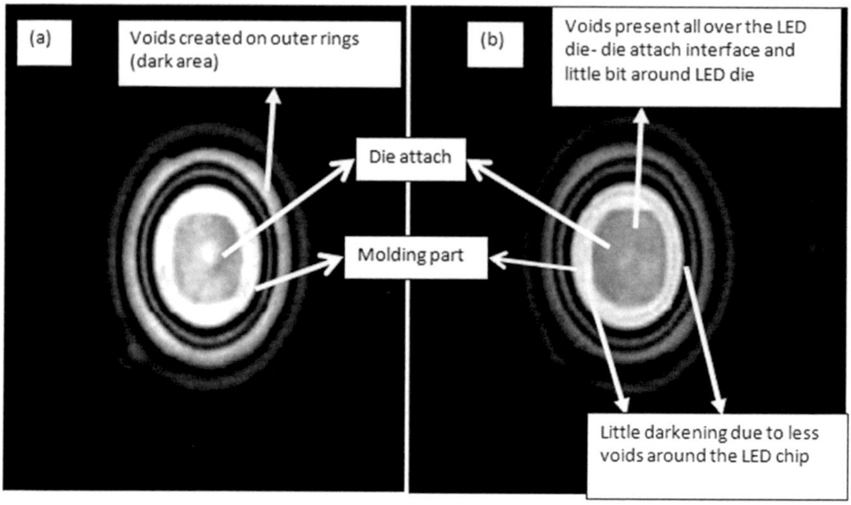


**Figure 7.12** CSAM images of the inverted blue LED (A) before initial degradation (tested 24 h) and (B) after final degradation (tested 356 h). The connection interface with the LED chip was completely dark after 356 h of testing, while the area outside the LED chip is not that dark. *(P. Singh, C.M. Tan, Degradation physics of high power LEDs in outdoor environment and the role of phosphor in the degradation process, Scientific Reports 6 (1) (2016), 1—13.* © Scientific Reports, *Nature. Reprinted with permission.)*

When LEDs are tested under various stress conditions, we can see the variations in their failure mechanisms. Under MET test, for example, LEDs experienced rapid deterioration in light output due to browning of the white molded silicone reflector part, as well as encapsulant detachment from the molding part, allowing moisture to infiltrate within the LED package and cause more damage to the LED. As a result, depending on the application of LEDs, different reliability tests should be conducted to assess their reliability under various operating conditions. For example, if a good heat sink is included, an electrical test alone might be appropriate for indoor LED use; if the heat sink is inadequate, an electrical-thermal test would suffice. MET, on the other hand, is needed for outdoor use, temperature and humidity factors may also be required during the durations when LEDs are not switched ON, such as in the daytime. When LEDs are turned ON in marine and automotive applications, they will be exposed to all three stress factors concurrently, which are high temperature, high moisture, and extreme electrical stresses, and an MET test is necessary. The MET test can aid the LED industry in better understanding the failure mechanisms associated with LEDs employed in extreme environments.

**Table 7.8** Failure modes and mechanisms associated with these single or multiple stresses on the LEDs.

| External stress (es) applied | Failure sites | Failure modes | Failure mechanisms | References |
|---|---|---|---|---|
| Electrical stress | Chip level | Lumen degradation, increase in reverse leakage current, increase in parasitic series resistance and short circuit | Defects in LED chip, electromigration of the metal atoms in the electrical contact to the surface of the LED die, instability in Mg diffusion in p-GaN layer and dislocation generation and movement at chip level | [17–27] |
| | Package level | Lumen degradation | Carbonization of encapsulation and phosphor thermal quenching | [28–34] |
| Thermal stress | Package level | Lumen degradation, detachment of encapsulant from molding part | Browning of the white reflector molding part of device | [35] |
| | Chip level | Lumen degradation | Degradation of the phosphors conversion efficiency | [35–37] |
| Electrical and thermal stress | Chip level | Lumen degradation | Crack in the LED chip | [21,22] |
| | Package level | Lumen degradation | Encapsulant yellowing, lens cracking or solder joint fatigue | [38–42] |

*Continued*

**Table 7.8** Failure modes and mechanisms associated with these single or multiple stresses on the LEDs.—cont'd

| External stress (es) applied | Failure sites | Failure modes | Failure mechanisms | References |
|---|---|---|---|---|
| Thermal and moisture stress | Chip level | Lumen degradation, forward voltage change, permanent destruction of LED i.e., no light | Wire ball bond fatigue | [6] |
| | Package level | Reduction of the overall light output | LED chip-die attach delamination, moisture entrapment in encapsulant, and lens cracking | [1−3,43,43] |
| Thermal, electrical and moisture stresses (MET test proposed in this work) | Chip level | Rapid lumen depreciation, forward voltage drop | LED chip/die attach delamination, defects in LED chip | [1] |
| | Package level | Rapid lumen degradation | Browning of the white silicone reflector molding part of device, encapsulant detachment from molding part | [1] |

P. Singh, C.M. Tan, Degradation physics of high power LEDs in outdoor environment and the role of phosphor in the degradation process, Scientific Reports 6 (1) (2016), 1−13. 

## 7.7 Summary

The combined effect of electric, moisture, and thermal stresses is used in this chapter to implement a new MET test. LEDs with and without phosphor layers produce different results. The key degradation reason in this test was found to be the gap produced at the molding–encapsulation interface due to large variation in thermo-expansivities coefficients between the two compounds. This gap allows moisture penetration in large amount inside the LED package, causing it to degrade rapidly. Moisture is also to blame for higher lumen loss and a decrease in Vf of LED I–V characteristics when they are turned on. Since the heat accumulated in white LEDs due to the phosphor evaporated the moisture from the LED die surface, phosphor presence helps in white LEDs protection. However, this heat from the phosphor will increase the degradation rate of LEDs while they are in use. As a result, phosphor-based white LEDs do not possess longer lifetime used in this study. RGB LEDs without phosphor that combine three colors to generate white light might be a better choice for future solid-state lighting than white LEDs, but no reliability research on such RGB white LEDs has been published.

When the failure mechanisms observed in the MET test were compared to those observed in other single or double stress variables, the importance of MET test can be understood. In practice, depending on the application, different reliability tests should be conducted to evaluate LED reliability, as the failure mechanisms associated with different tests can be very different.

## References

[1] P. Singh, C.M. Tan, Degradation physics of high power LEDs in outdoor environment and the role of phosphor in the degradation process, Scientific Reports 6 (1) (2016) 1–13.

[2] S.H. Chen, C.M. Tan, B.K.E. Chen, Ensuring accuracy in optical and electrical measurement of ultra-bright LEDs during reliability test, Microelectronics Reliability 52 (2012) 1632–1635.

[3] J.J. Fan, K.C. Yung, M. Pecht, Lifetime estimation of high-power white LED using degradation-data-driven method, IEEE Transactions on Device and Materials Reliability 12 (2012) 470–477, https://doi.org/10.1109/Tdmr.2012.2190415.

[4] M. Arik, et al., Effects of localized heat generations due to the color conversion in phosphor particles and layers of high brightness light emitting diodes, Proceedings of the ASME International Electronic Packaging Technical Conference (2003) 611–619, https://doi.org/10.1115/IPACK2003-35015.

[5] D.H. Yoon, et al., Thermally stable phosphor-in-glass for enhancement of characteristic in high power LED applications, Materials Letters 157 (October 15 , 2015) 89–92.

[6] C.M. Tan, P. Singh, Time evolution degradation physics in high power white LEDs under high temperature-humidity conditions, IEEE Transactions on Device and Materials Reliability 14 (2014) 742–750.

[7] I.H. Lin, et al., Chemical Compatibility of LEDs, Application Note. Technical Report, 2015. Available at: http://www.osram-os.com/Graphics/XPic0/00161697_0.pdf/Chemical%20Compatibility%20of%20LEDpdf (Accessed: 30th August 2015).

[8] J.Z. Hu, L.Q. Yang, W.J. Hwang, M.W. Shin, Thermal and mechanical analysis of delamination in GaN-based light-emitting diode packages, Journal of Crystal Growth 288 (2006) 157–161.

[9] J.Z. Hu, L.Q. Yang, M.W. Shin, Thermal and mechanical analysis of high-power LEDs with ceramic packages, IEEE Transactions on Device and Materials Reliability 8 (2008) 297–303.
[10] OSRAM Opto Semiconductors, Golden DRAGON Datasheet Version 1.0 Technical report, 2015. Available at: http://www.osram-os.com/Graphics/XPic5/00179585_0.pdf/LW%20W5SM.pdf (Accessed: 30th August 2015).
[11] L. Kim, M.W. Shin, Thermal resistance measurement of LED package with multichips, IEEE Transactions on Components and Packaging Technologies 30 (4) (2007) 632–636.
[12] B. Pardo, A. Gasse, A. Fargeix, J. Jakovenko, R.J. Werkhoven, X. Perpiñà, P. Bancken, Thermal resistance investigations on new leadframe-based LED packages and boards, Microelectronics Reliability 53 (8) (2013) 1084–1094.
[13] P. Zeng, et al., Hygro-thermo-mechanical modeling of LED luminaires, in: 14th International Conference on Electronic Packaging Technology (ICEPT) 606–611, 2013, https://doi.org/10.1109/ICEPT.2013.6756543.
[14] M. Zachau, et al., Phosphors for solid state lighting, SPIE Proceedings (2008), https://doi.org/10.1117/12.760066.
[15] P.F. Smet, A.B. Parmentier, D. Poelman, Selecting conversion phosphors for white light-emitting diodes, Journal of the Electrochemical Society 158 (2011) R37–R54, https://doi.org/10.1149/1.3568524.
[16] R. Hu, X.B. Luo, H. Zheng, Hotspot location shift in the high-power phosphor-converted white light-emitting diode packages, Japanese Journal of Applied Physics 51 (2012), https://doi.org/doi: Artn 09mk0510.1143/Jjap.51.09mk05.
[17] J. Arnold, When the light go out: LED failure mode and mechanisms Technical report, Available at: http://www.dfrsolutions.com/publications/when-the-lights-go-out-led-failure-modes-and-mechanisms/, 2008 (Accessed: 30th August 2015).
[18] M. Pavesi, et al., Optical evidence of an electrothermal degradation of InGaN-based light-emitting diodes during electrical stress, Applied Physics Letters 84 (2004) 3403–3405, https://doi.org/10.1063/1.1734682.
[19] G. Meneghesso, et al., Failure mechanisms of GaN-based LEDs related with instabilities in doping profile and deep levels, in: Reliability Physics Symposium Proceedings, 2004. 42nd Annual. 2004 IEEE International 474–478, 25–29 April 2004, https://doi.org/10.1109/RELPHY.2004.1315374.
[20] P. Kozodoy, S.P. DenBaars, U.K. Mishra, Depletion region effects in Mg-doped GaN, Journal of Applied Physics 87 (2000) 770–775, https://doi.org/10.1063/1.371939.
[21] M. Meneghini, et al., Failure mechanisms of gallium nitride LEDs related with passivation, IEEE IEDM Technical Digest (5-5 Dec. 2005), https://doi.org/10.1109/IEDM.2005.1609534.1009–12.
[22] G. Lu, S. Yang, Y. Huang, Analysis on failure modes and mechanisms of LED, in: 8th International Conference on Reliability, Maintainability and Safety, 2009. ICRMS 2009, 20–24 July 2009, pp. 1237–1241, https://doi.org/10.1109/ICRMS.2009.5270043.
[23] G. Meneghesso, et al., Failure modes and mechanisms of DC-aged GaN LEDs, Physica Status Solidi A 194 (2002) 389–392.
[24] G. Meneghesso, et al., Degradation mechanisms of GaN-based LEDs after accelerated DC current aging, in: International Electron Devices Meeting, 2002. IEDM '02 103–106, 8–11 Dec. 2002, https://doi.org/10.1109/IEDM.2002.1175789.
[25] M. Pavesi, F. Rossi, E. Zanoni, Effects of extreme dc-ageing and electron-beam irradiation in InGaN/AlGaN/GaN light-emitting diodes, Semiconductor Science and Technology 21 (2006) 138–143.
[26] M. Ott, Capabilities and reliability of LEDs and laser diodes, International NASA Parts and Packaging Publication 20 (6) (1996) 1–7.

[27] S.L. Chuang, et al., Kinetic model for degradation of light-emitting diodes, IEEE Journal of Quantum Electronics 33 (1997) 970–979.
[28] P. McCluskey, et al., Reliability of commercial plastic encapsulated microelectronics at temperatures from 125°C to 300°C, in: The Third European Conference on High Temperature Electronics, 1999. HITEN 99, 155–162, 04 Jul 1999-07 Jul 1999, https://doi.org/10.1109/HITEN.1999.827481.
[29] P. McCluskey, K. Mensah, C. O'Connor, A. Gallo, Reliable use of commercial technology in high temperature environments, Microelectronics Reliability 40 (2000) 1671–1678, https://doi.org/10.1016/S0026-2714(00)00188-8.
[30] G. Meneghesso, et al., Reliability of visible GaN LEDs in plastic package, Microelectronics Reliability 43 (2003) 1737–1742, https://doi.org/10.1016/S0026-2714(03)00289-0.
[31] M. Meneghini, et al., High temperature electro-optical degradation of InGaN/GaN HBLEDs, Microelectronics Reliability 47 (2007) 1625–1629, https://doi.org/10.1016/j.microrel.2007.07.081.
[32] F. Wu, Y. Wu, B. An, F. Wu, Analysis of dark stain on chip surface of high-power LED, in: 7th International Conference on Electronic Packaging Technology. 1–4, 26–29 Aug. 2006, https://doi.org/10.1109/ICEPT.2006.359785.
[33] L. Zhou, B. An, Y. Wu, S. Liu, Analysis of delamination and darkening in high power LED packaging, in: 16th IEEE International Symposium on the Physical and Failure Analysis of Integrated Circuits, 2009. IPFA 2009 656–660, 6–10 July 2009, https://doi.org/10.1109/IPFA.2009.5232549.
[34] R.J. Xie, N. Hirosaki, N. Kimura, K. Sakuma, M. Mitomo, 2-phosphor-converted white light-emitting diodes using oxynitride/nitride phosphors, Applied Physics Letters 90 (2007) doi: Artn 19110110.1063/1.2737375.
[35] L.R. Trevisanello, et al., Thermal stability analysis of high brightness LED during high temperature and electrical aging, in: Proceedings of the SPIE 6669, Seventh International Conference on Solid State Lighting, vol. 6669, September 14, 2007, https://doi.org/10.1117/12.732398.
[36] N. Trivellin, et al., Degradation mechanisms of high-power LEDs for lighting applications: an overview, IEEE Transactions on Industry Applications 50 (1) (2014) 78–85.
[37] J.L. Down, The yellowing of epoxy resin adhesives: report on high-intensity light aging, Studies in Conservation 31 (1986) 159–170.
[38] Q. Zhang, et al., Dynamic mechanical properties of the transparent silicone resin for high power LED packaging, in: International Conference on Electronic Packaging Technology & High Density Packaging, 2008. ICEPT-HDP 2008. 1–4, 28–31 July 2008, https://doi.org/10.1109/ICEPT.2008.4607073.
[39] M. Meneghini, L.R. Trevisanello, G. Meneghesso, E. Zanoni, A review on the reliability of GaN-based LEDs, IEEE Transactions on Device and Materials Reliability 8 (2008) 323–331, https://doi.org/10.1109/Tdmr.2008.921527.
[40] M.H. Chang, D. Das, M. Pecht, Interconnect Reliability Assessment of High Power Light Emitting Diodes (LEDs) through Simulation 2nd International Conference on Reliability, Safety and Hazard (ICRESH), 2010, pp. 418–424, https://doi.org/10.1109/ICRESH.2010.5779586 (14-16 Dec. 2010).
[41] D.S. Kim, B. Han, Y.-J. Kim, Degradation analysis of secondary lens system and its effect on performance of LED-based luminaire, Microelectronics Reliability 54 (1) (2014) 131–137.
[42] A. Gladkov, A. Bar-Cohen, Parametric dependence of fatigue of electronic adhesives, IEEE Transactions on Components and Packaging Technologies 22 (1999) 200–208.
[43] H.H. Kim, et al., Thermal transient characteristics of die attach in high power LED PKG, Microelectronics Reliability 48 (2008) 445–454, https://doi.org/10.1016/j.microrel.2007.08.009.

# CHAPTER 8

# Packaged LEDs degradation under UV radiation effect

## 8.1 Introduction

Effect of various environmental factors on the LEDs lifetime is studied in previous chapters. Ultraviolet (UV) radiation is another important environmental factor that can reduce the LEDs lifetime due to its higher energy effecting the various parts of a packaged LED [1,2]. However, most works discussed the impact of temperature, humidity, or electrical input on LEDs light output degradation and very rarely the effect of UV radiation on LEDs parameters was discussed. UV radiation is a form of nonionizing radiation that is emitted by sun, and it can impact LEDs either on the earth's surface or in the outer surface [3,4]. However, the impact would be severe in the outer space as lower-wavelength UV rays, which are absorbed by ozone, are present there [5,6]. Higher-wavelength UV rays, which are not absorbed by ozone layer, and they are generally not considered harmful for humans will impact the electronic products on the earth. These UV radiations reaching the earth surface are generally in the range of 315−399 nm (also known as UV-A) will be the topic of interest in this chapter.

There are few studies discussing the effect of UV radiation on silicone material, which is the most common material used in the LED packaging as discussed in the earlier Chapters. Ozcam et al. [7] performed a study to observe the effect of UV/ozone on silicone elastomers (Poly dimethyl siloxanes, PDMS and Poly Vinyl methyl siloxanes, PVMS). Their work reveals that PVMS and PDMS underwent change in their properties when exposed under UV radiation. Specifically, a large number of hydrophilic groups were observed due to the interactions with reactive atomic oxygen and UV radiation in both materials. PVMS exhibits change in properties at a faster rate when compared with PDMS. They also found that PVMS underwent dramatic changes throughout the entire bulk while PDMS was modified primarily close to the surface under UV radiation. They also

*Reliability and Failure Analysis of High-Power LED Packaging*
ISBN 978-0-12-822408-3
https://doi.org/10.1016/B978-0-12-822408-3.00001-0

reported that under UV exposure, the top surface, which is around 5 nm of PDMS, gets converted into a silica-like material, which later acts as a resistance for oxygen atom diffusing into the surface. However, in their case, the PDMS bulk maintains its elasticity. In contrast, both PVMS film's surface and bulk underwent substantial changes in properties when exposed to UV. First, the surface modification of PVMS takes place after initial few seconds of the UV radiation. In addition, they also experienced substantial bulk properties modification, which includes complete densification along with increase in bulk modulus. The susceptibility of the vinyl bonds to radial reactions was found to be responsible for this effect.

Similarly, Fischer et al [8] also investigated the effect of UV radiation on silicone glues, which are generally used as an encapsulant in photovoltaic devices. Silicone underwent discoloration and embrittlement after UV radiation. Fig. 8.1 shows discoloration for three different types of silicones employed in their work.

They connected these failures with radical generation upon UV radiation. Fischer at al. believed that after radical formation, cross-linking, and thereafter cracking and embrittlement occurred due to shrinkage. Polymers degradation via a radical mechanism can lead to the formation of small metal clusters, which contain residues of the silicone and Karstedt catalyst. Both types of nanoparticles contribute to a discoloration of the silicone glues. The use of ultrapure starting substances is a possible method for avoiding this, with catalyst amount to a minimum and a denser cross-link density of the silicone that minimizes the mobility of dangling chains formed when exposed to UV and their tendency of forming metal clusters.

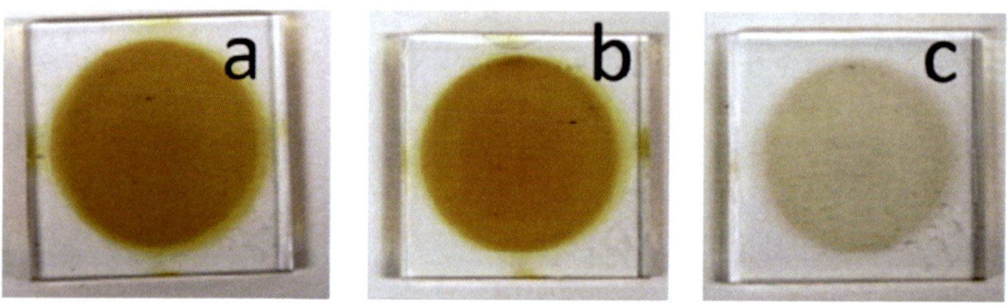


**Figure 8.1** Micrographs of the silicone glues sandwiched samples after exposure from UV radiation under vacuum using the ESTEC SUV facility, (A) Elastosil S 690, (B) Elastosil S 695, (c) Dow DC 93-500, dimensions 19 × 19 mm, thickness 25 microns [8]. *(N. Narendran, Y. Gu, J.P. Freyssinier, H. Yu, L. Deng, Solid-state lighting: failure analysis of white LEDs, Journal of Crystal Growth 268 (3–4) (2004) 449–456.* © Polymer Degradation and Stability, *Elsevier. Reprinted with permission.)*

These abovementioned works demonstrated the effect of UV radiation on LEDs package material, but the impact on overall LEDs reliability is yet to be studied, especially with and without phosphor.

In this chapter, the effect of UV radiation on LED's reliability will be investigated. A test was designed for this purpose to explore the degradation physics, and both blue and white LEDs were used to undergo the MET test.

## 8.2 Experimentation

Two sets of high-power OSRAM golden dragon LEDs, each consisting of three units of white and blue LEDS respectively, were chosen in this work. All the LEDs were powered ON with 350 mA constant current, according to manufacturing specifications. During the test, LEDs were placed in the UV chamber from T-MACHINE model T-07-180201 for accelerated life test with UV-A radiation along with 85°C/85% RH to mimic the real-time scenario for the LEDs.

The electrical measurements for all the LEDs were done using Keithley source meter model 2651A. Optical measurements were done with integrating sphere model Mentor's TeraLED. Initial set of measurements were done on all of them to serve as reference baseline for each test sample.

The samples were removed from the UV radiation chamber every 24 h for their electrical and optical measurements. The measurement settings for all were according to Tan et al. [9,10]. All the measurements were done within 3 h from the time they were taken out from the test chamber in order to reduce the probability of moisture evaporation from the LED package. Optical microscope was used to examine any possible visual defects of the LED packages after the tests, and SEM/EDS was employed to observe the possible elemental composition change for the package material during the test. Electrical characterization of these tested LEDs under different conditions was also compared.

## 8.3 Experimental results

The lumen degradation percentages versus time for white and blue LEDs tested under power ON condition are shown in Fig. 8.2.

LEDs were tested until they reach 20% lumen degradation after which they were treated as failed. It is interesting to observe that blue LEDs degraded at a higher rate when compared with white LEDs during this test.

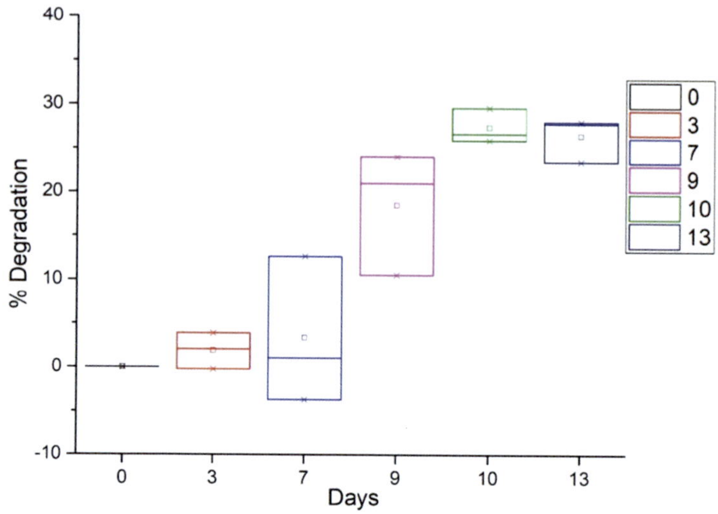


(a)

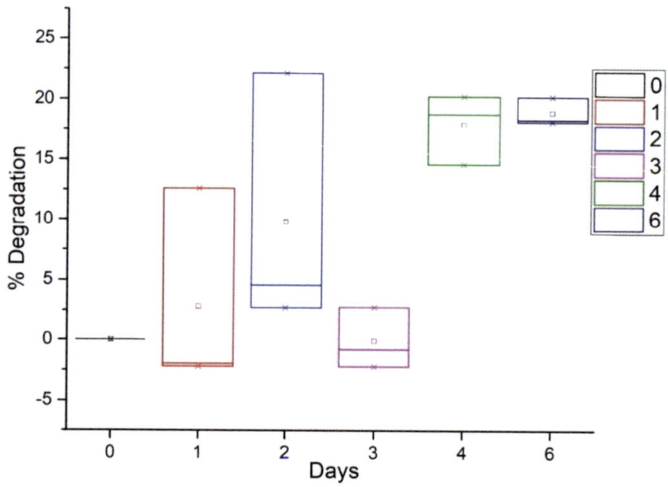


(b)

**Figure 8.2** Percentage lumen degradation variation for (A) white and (B) blue LEDs under UV radiation.

In general, during the temperature-humidity tests, white LEDs degraded faster than blue LEDs due to phosphor heat generation as discussed in the previous chapter.

I–V curves were also measured for the LEDs during the test for blue and white LEDs under test as shown in Fig. 8.3.

Negligible change in the forward voltage is observed during the test for both blue and white LEDs, which indicates that the LED chips were intact during the test and the lumen degradation was not due to chip failure. Blue-to-Yellow Ratio (BYR) is calculated for the white LEDs, and no change was observed, which shows that phosphor was also in good health. LED package degradation should therefore be the reason that led to lumen degradation.

## 8.4 Physical analysis on the test samples

Optical microscope (model KEYENCE 5000x) was employed to observe the LED package during the UV radiation test. Optical images of the blue

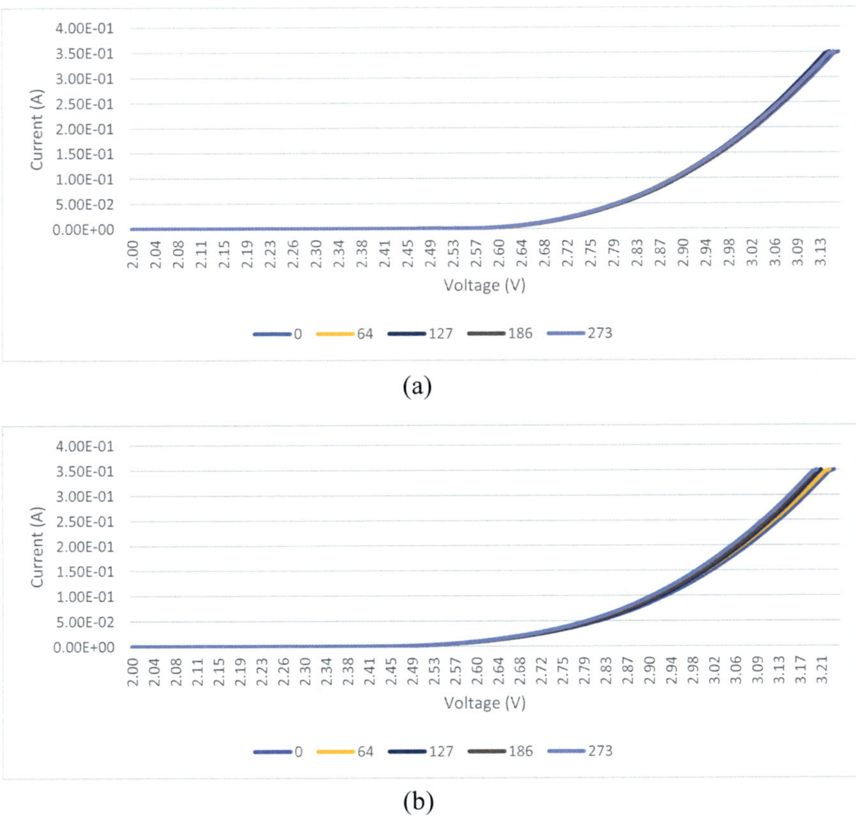


**Figure 8.3** I–V curves of (A) white and (B) blue LEDs tested under UV radiation.

and white LEDs under test are shown in Fig. 8.4. Clear discoloration was observed for both the LEDs; however, the time for the start of degradation was different for the two sets under study.

Blue LEDs package exhibited discoloration after 21 h, whereas white LEDs exhibited discoloration after 147 h. This is a very unique trend because during the previous studies, white LEDs show rapid discoloration when compared with blue LEDs. It also indicates that the effect of light is dominant when compared with temperature or moisture because under similar test conditions and without UV radiation, LEDs experienced slower degradation rate as reported in previous work.

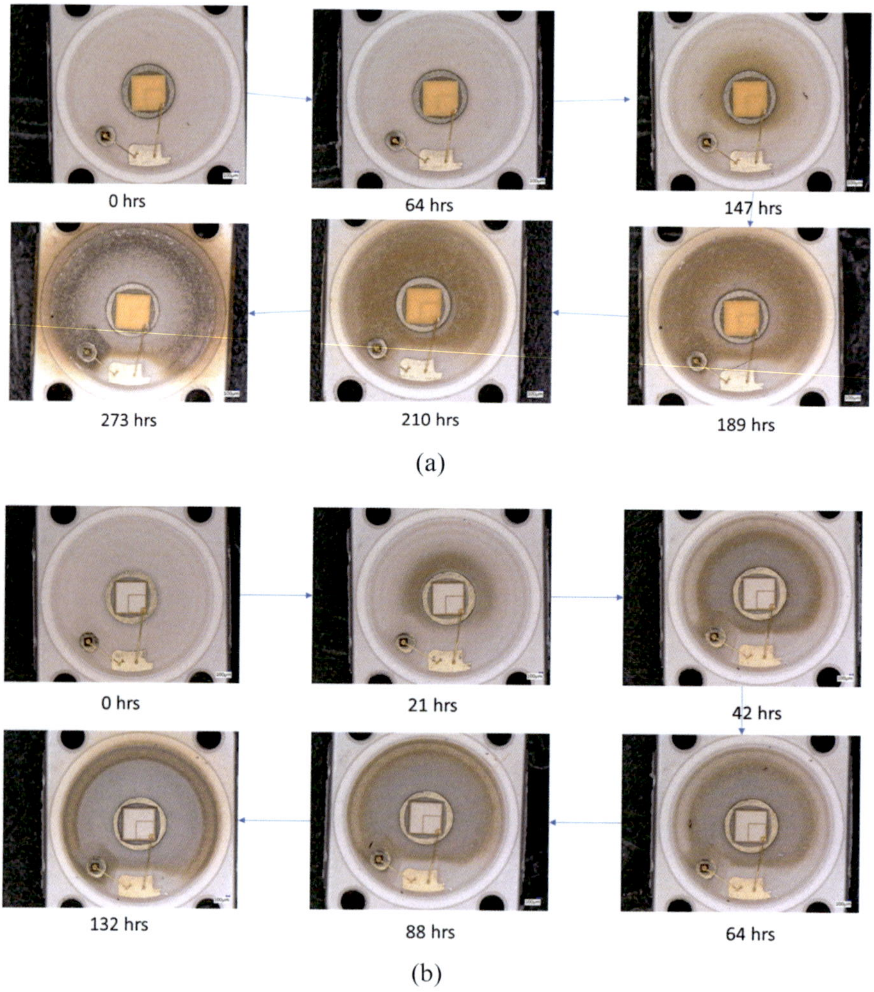


**Figure 8.4** Optical microscope images of (A) white and (B) blue LEDs when taken at 50x magnification.

In these experiments, there was external UV light source and hence degradation was expected to occur in both encapsulant and molding part for the LEDs under test. In case of blue LEDs, there is a presence of shorter wavelength, i.e., combination of blue and UV light, which has higher energy and can therefore cause severe package degradation. For the case of white LED, most of the blue LED light is converted to yellow to produce white light and thus the intensity of the short wavelength high energy source is lesser in comparison to that of the blue LEDs, and this is believed to be the reason for slower white LED degradation compared to blue LED under UV radiation.

To observe package degradation for the LEDs, encapsulant was detached from the LED package and LED's encapsulant and molding part were more closely examined using higher magnification optical microscope. Optical images of blue and white LEDs molding and encapsulant are shown in Fig. 8.5.

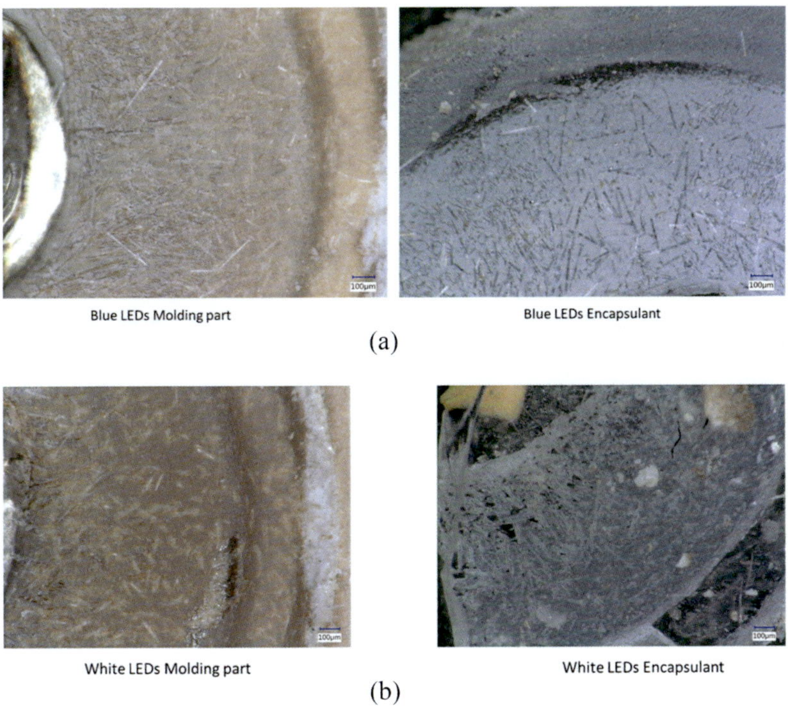


**Figure 8.5** Optical micrographs of (A) Blue and (B) White LEDs molding part and encapsulant taken at 1000x magnification.

Degradation was observed for both encapsulant and molding part of LEDs, which was not the case when LEDs were tested under high temperature and moisture conditions without UV light. Also, blue LEDs encapsulant and molding part showed higher discoloration and cracks compared with white LEDs as observed in Fig. 8.5. This further supports the postulation that shorter-wavelength light is indeed a more dominant factor for LEDs lumen degradation when compared with temperature and moisture. SEM was employed to observe the cracks on the molding and encapsulant surface for both the LEDs, which can provide more clear indication of the degradation.

SEM images from Fig. 8.6 show that more filler-like particles were observed for blue LEDs when compared with white LEDs. These filler-like particles were also observed by Singh et al. [11,12] in their previous studies

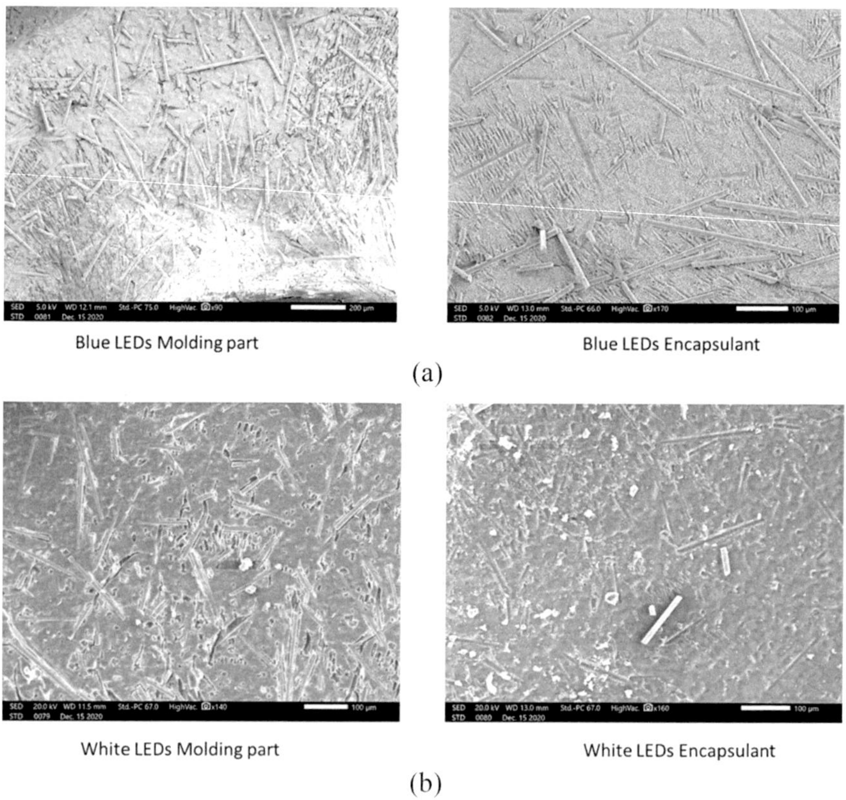


**Figure 8.6** SEM images of blue and white LEDs under UV radiation.

and silicone oxidation was found to be the cause. Under high temperature and moisture conditions along with light, silicone degrades and can undergo various degradation mechanisms. As discussed earlier, moisture will be higher inside the blue LEDs package due to lower inner temperature compared to white LEDs, and blue LEDs will have higher amount of high energy light (blue + UV light), and therefore, higher oxidation was expected in blue LEDs. EDAX was employed to verify the oxidation trend for silicone in both the LEDs.

From Tables 8.1 and 8.2, it was observed that oxygen content was higher for blue LEDs molding part after the UV radiation test. The difference in oxygen content between blue and white LEDs molding part was found to be same at different acceleration voltages (5, 10, and 15 keV), which shows that the molding was degraded from surface to bulk and also the degradation mechanism has evolved during the test time, otherwise higher degradation will be expected at the surface than at the bulk, which was not the case. Similarly, EDAX was employed to observe the elemental composition of blue and white LED's encapsulant. Tables 8.3 and 8.4 shows the EDAX results for encapsulant.

The difference in oxygen content here was more significant at the encapsulant site between blue and white LEDs, which was expected because UV mostly incident on the encapsulant. Molding is situated in the inner side of the package and should be less affected. Along with higher amount of shorter-wavelength light, blue LEDs also have higher moisture penetration compared to white LEDs as observed from previous studies, and this higher amount of moisture could lead to higher rate of hydrolysis that will cause oxidation to occur earlier at the end for the blue LEDs. The

**Table 8.1** Elemental composition of blue LED molding part at an acceleration voltage of 15 keV.

| Element | Weight % |
|---|---|
| O | 37.6 |
| C | 24.6 |
| Ti | 18.9 |
| Si | 8.6 |
| Pt | 4.1 |
| Ca | 4.0 |
| Al | 1.9 |
| Mg | 0.2 |
| Na | 0.2 |

**Table 8.2** Elemental composition of white LED molding part at an acceleration voltage of 15 keV.

| Element | Weight % |
|---|---|
| O | 35.2 |
| Ti | 24.1 |
| C | 21.8 |
| Si | 8.9 |
| Pt | 5.2 |
| Ca | 3.1 |
| Al | 1.6 |
| Mg | 0.2 |

**Table 8.3** Elemental composition of blue LED encapsulant at an acceleration voltage of 15 keV.

| Element | Weight % |
|---|---|
| O | 38.3 |
| Si | 18.8 |
| Ti | 18.5 |
| C | 17.4 |
| Pt | 3.3 |
| Ca | 1.9 |
| Al | 1.4 |
| Na | 0.2 |
| Mg | 0.2 |

**Table 8.4** Elemental composition of white LED encapsulant at an acceleration voltage of 15 keV.

| Element | Weight % |
|---|---|
| Si | 33.5 |
| C | 27.4 |
| O | 25.5 |
| Ti | 8.3 |
| Pt | 4.7 |
| Al | 0.3 |
| Ca | 0.2 |

processes of silicone degradation at high temperature, high humidity, and light have been investigated to be hydrolysis, condensation, oxidation, and thermal aging as discussed in Chapter 7. These accelerated tests were performed under UV radiation along with high temperature and moisture, and thus, moisture can penetrate deeper inside the LED package due to various mechanisms as investigated by Singh et al. [11,12] in their work.

The above studies deal with the silicone material degradation under the UV radiation; however, none of studies reveal the effect of such silicone degradation on the overall lumen degradation for the LEDs. This work shows the unique phenomenon where blue LEDs degrade at a rapid and faster rate when compared to white LEDs, which was not expected earlier. More time-dependent studies are needed in this field to understand the silicone failure mechanisms in such harsh environments.

## 8.5 Summary

This chapter introduces the effect of UV radiation on high-power blue and white LEDs along with the common stressed factors such as temperature, moisture, and light. UV radiation effect on silicone material was studied before but its effect on the overall LED lumen was not investigated. The experiment reported here is a first of its kind and unique degradation trends are observed for the LEDs. Blue LEDs underwent rapid lumen degradation with higher degradation rate than white LEDs. Degradation of LEDs package including molding part and encapsulant was found to be the cause for such LED lumen degradation as chip and phosphor were found to be intact. Microscopic examinations showed cracks and discoloration at both encapsulant and molding part of LEDs, with more cracks and filler-like particles on blue LEDs molding and encapsulant surface as a result of oxidation. Higher amount of short-wavelength light, i.e., blue and UV light in Blue LEDs, is found to be the reason for their rapid degradation because of the combined high energy light sources, together with the presence of moisture, causing rapid hydrolysis that leads to earlier oxidation reaction, and consequently, higher and faster lumen degradation was observed for blue LEDs in this work.

## References

[1] N. Narendran, Y. Gu, J.P. Freyssinier, H. Yu, L. Deng, Solid-state lighting: failure analysis of white LEDs, Journal of Crystal Growth 268 (3–4) (2004) 449–456.

[2] M. Yazdan Mehr, W. Van Driel, F. De Buyl, K. Zhang, Study on the degradation of optical silicone exposed to harsh environments, Materials 11 (8) (2018) 1305.
[3] F.R. De Gruijl, Skin cancer and solar UV radiation, European Journal of Cancer 35 (14) (1999) 2003–2009.
[4] N.D. Paul, D. Gwynn-Jones, Ecological roles of solar UV radiation: towards an integrated approach, Trends in Ecology & Evolution 18 (1) (2003) 48–55.
[5] C.S. Cockell, Biological effects of high ultraviolet radiation on early Earth—a theoretical evaluation, Journal of Theoretical Biology 193 (4) (1998) 717–729.
[6] B.L. Diffey, Sources and measurement of ultraviolet radiation, Methods 28 (1) (2002) 4–13.
[7] A.E. Özçam, K. Efimenko, J. Genzer, Effect of ultraviolet/ozone treatment on the surface and bulk properties of poly (dimethyl siloxane) and poly (vinylmethyl siloxane) networks, Polymer 55 (14) (2014) 3107–3119.
[8] H.R. Fischer, C. Semprimoschnig, C. Mooney, T. Rohr, E.R. van Eck, M.H. Verkuijlen, Degradation mechanism of silicone glues under UV irradiation and options for designing materials with increased stability, Polymer Degradation and Stability 98 (3) (2013) 720–726.
[9] P. Singh, C.M. Tan, L.B. Chang, Early degradation of high power packaged LEDs under humid conditions and its recovery—myth of reliability rejuvenation, Microelectronics Reliability 61 (2016) 145–153.
[10] C.M. Tan, B.K. Chen, X. Li, S.J. Chen, Rapid light output degradation of GaN-based packaged LED in the early stage of humidity test, IEEE Transactions on Device and Materials Reliability 12 (1) (2011) 44–48.
[11] P. Singh, C.M. Tan, Degradation physics of high power LEDs in outdoor environment and the role of phosphor in the degradation process, Scientific Reports 6 (1) (2016) 1–13.
[12] P. Singh, C.M. Tan, Time evolution of packaged LED lamp degradation in outdoor applications, Optical Materials 86 (2018) 148–154.

CHAPTER 9

# Reliability tests standards for LEDs

## 9.1 Introduction

As observed from previous chapters, light-emitting diodes (LEDs) provide numerous advantages and exceptional reliability, due either to its robustness, which is essential for automotive industry, or longer lifetime compared to its counterparts, which is important for general lighting. In the past decade, reliability tests at component level were the focus as can be seen from the development of various standards, namely LED lumen maintenance test (ANSI/IES LM-79 or LM-80) [1,2], lumen maintenance projection (IES TM-21) [3—5], catastrophic failure rate projection (IES TM-26) [6—8], and driver robustness test (ANSI C82.16) [9].

Most reputable LED manufacturers conduct reliability tests internally to ensure good reliability of their products [10]. The design for these tests is often based on process experiences, product development, and field quality data, rendering the selection of test parameters confidential or trade secret for each manufacturer in conducting the tests [10]. On the other hand, regulatory and market-transformation organizations such as Design Lights Consortium (DLC) [11,12], Energy Star [13], the California Energy Commission [14,15], etc., would like to have standardized approaches to allow comparisons for LED lamp and luminaire-level reliability measurements, including performance measurements, such as lumen or color maintenance and catastrophic failure rate. Apart from these organizations, specifiers, lighting designers and consumers also possess the same desire for fair comparison purposes.

However, there is a lack of widely accepted LED lighting reliability standards for the industry, which can be employed to test the reliability of LEDs under all stresses that an LED can encounter in real-time applications. Singh and Tan (as shown in earlier chapters) proposed a new MET test in this regard, but it yet to gain wide acceptance. Singh and Tan [16] also

*Reliability and Failure Analysis of High-Power LED Packaging*
ISBN 978-0-12-822408-3
https://doi.org/10.1016/B978-0-12-822408-3.00007-1

showed the evolving nature of the degradation mechanism for LEDs at different time and under different stress conditions, which demands a better extrapolation technique [17] to determine LEDs reliability at normal operating conditions from accelerated test data.

The importance of reliability test standards, the current reliability test standards, the difficulties in extrapolating reliability data to normal operating conditions, and the need for additional reliability test standards will be discussed in this chapter.

## 9.2 Necessity for reliability test standards

The fact that a single failure mode can be caused by multiple failure mechanisms [18] is one of the major challenges in the LED reliability report. For example, the deterioration of different LED components, such as the LED chip, lens, and phosphor, can affect the luminous flux degradation, and they can be used as key failure criterions for LEDs [19]. In other words, a single LED should be considered as a system that consists of different parts, namely LED chip, electrical connections, housing, heat sink, die attachment, and encapsulation. The deterioration of each component, as well as their interactions, can have an effect on LED's overall reliability. In order for luminaire manufacturers to make lifetime predictions, the reliability of an LED is a combination of the reliability of different components. From the standpoint of an LED maker, having access to individual contributions from parts in an LED is critical for improving the LED package's reliability. In fact, this information will also help in identify the weak parts in the component so that research effort can focus on these parts to increase the overall LED reliability. However, to understand or obtain such information, new reliability tests are required because LED reliability test standards today do not emphasize the need on this aspect of LED reliability.

In the past few years, study of the aging of the parts individually, such as phosphor aging or silicone aging, was performed. The idea behind these aging tests is that these can bring information regarding the different aging rates and can help to evaluate the overall LED reliability. However, these separate aging tests might not lead to optimistic results because these results do not consider the interaction between various parts [18]. For example, LED's encapsulant and molding could be degraded due to various stress factors such as the blue light from LED die as well as yellow light from phosphor. Their degradation can also induce the aging of other parts such as die attachment, etc.

Due to the abovementioned issues, there is a need for better LED testing standards to have more reliable testing and accurate lifetime estimation of LEDs. However, before getting into the need for new extrapolation techniques and better reliability test standards, we would like the readers to know the present LED reliability test standards, which will be discussed in the next section.

## 9.3 Current LED reliability test standards

First initiative in developing LED reliability test standards was taken by the International Electrotechnical Commission (IEC) [20]. A project (NP) [21] was initiated in the IEC to develop a performance standard that established a method and criteria for optimizing test duration of LED-based products for general lighting, making use of principal component reliability testing in September 2012. The IEC project planned to include principal components such as optical materials, LED package and interconnects, cooling systems, electronics, mechanical structure, etc.

After 4 years of developments, IEC established a draft with the following scope: "The technical specification outlines how to use principal component testing of LED light sources and LED luminaires for general lighting to determine product reliability. Initial certification tests and accelerated stress tests of principal components are among the methods and standards used. The performance of any major component would have an effect on the final product's performance. LED performance standards currently define reliability requirements for a restricted period of operation (e.g., 6000 h). For various LED product types (e.g., LED lamp, LED module, and LED luminaire) the relevant product performance standard will detail the full extent of the reliability assessment that is necessary. This technical specification does not cover techniques for validating complete lifetime statements or projecting lumen maintenance."

IEC document had two main philosophies that constitute the pillar stones of this document. One philosophy was based on testing the principal components. If these components pass the test, then after all the components are assembled, the final lighting product should be somewhat more reliable. The second philosophy was focused on accelerated tests achieved by overstressing the tested components. This philosophy assumes that if the components are tested in overstressed conditions for a short period of time, then the components should survive for a much longer time period when they are used under normal conditions.

IEC reliability standard identifies the first principal component as the LED package. The document is written to evaluate if an LED package is capable of passing the specified stress tests. The stress tests are selected so that if the LED package passes the tests, it can be expected to give a certain level of reliability in general lighting applications.

The IEC standard lists the initial LED package tests that are largely based on JEDEC standards. They are grouped into:

**(1)** low-temperature operating life (LTOL) test [22], temperature and operation stress, including temperature and operation stress test, pulsed operating life (PLT) test [23], high-temperature operating life (HTOL) test [24,25];

**(2)** vibration variable frequency (VVF) test, thermomechanical stress, including temperature cycling (TMCL) test [26,27];

**(3)** temperature and humidity stress, including damp heat cycling (DHC) test and wet high-temperature operating life (WHTOL) test [26,27];

**(4)** electrical stress, including discharge with machine model (ESD-MM) test and electrostatic discharge with human body model (ESD-HBM) test, [26,27]; and

**(5)** environmental stress, including flowing mixed gas corrosion (FMGC) test, hydrogen sulfide ($H_2S$) test, and sulfur dioxide test ($SO_2$) test [26,27].

The VVF test is the first test conducted on LED interconnect. IEC also introduces temperature-cycling interconnect test (TMCL) as an accelerated stress test (AST) with the purpose to demonstrate the integrity and reliability of the solder joint against mechanical stress induced by severe temperature changes.

The IEC standard then considers the second principal component, which is the optical material. Optical materials include any transmissive or reflective materials and converting material, such as remote phosphors or color filters used in LED lighting products. A relative humidity (RH) test, boiling water (BW) test, oven water (OW) test, and high-temperature exposure (HTE) test are the first tests for optical materials. The temperature/humidity (TH) and temperature/light exposure (TL) measurements are accelerated stress tests for optical materials. The TH accelerated stress test evaluates the performance of the optical material under stresses caused by moisture and temperature, while the TL accelerated stress test evaluates the performance of the optical material under stresses caused by temperature and light exposure.

The next principal component in the IEC standard is the electronic subassembly. The initial tests include temperature/operation stress (PTC) and humidity/operation stress (HOT) test.

The IEC standard continues to list the tests for other components, such as cooling system and construction materials. For each principal component, the IEC standard provides the failure criteria. For example, the LED packages criteria include the changes of voltage and color to certain extents before and after the initial tests.

IEC approach for the final product level test is to integrate the test data from the component tests and apply them to the final products, with the aid of experiences on key components in the final assembly.

### 9.3.1 Illumination engineering society (IES) standards

The IES reliability standard development took an approach similar to the IEC [28–30]. The first draft of IES LM-87 focuses on the LED package level reliability or robustness tests that are under the overstress test conditions. These overstress test methods are provided in the JEDEC standards. The IES document defines failure, failure criteria, and LED package-specific test conditions. It identifies six overstress tests that reflect general lighting applications: HTOL, TMCL, resistance to soldering heat (RSH), WHTOL, electrostatic discharge, and continuous switching cycles until failure tests. This group of tests is based on current industry best practices among LED manufacturers.

Although the results may not be directly interpreted as a quantitative measure of LED product reliability, they can indicate the robustness of the LED package. In turn, they help manufacturers to set and confirm product performance specifications. Parallel to the LED package-level reliability test development, IES also proposed to develop a document for LED lamp and luminaire level tests.

### 9.3.2 ASSIST project

Choosing the critical variables to test LED lighting product reliability is essential for establishing standardized test methods and a possible projection method using test data. The Rensselaer Polytechnic Institute Lighting Research Center ASSIST (Alliance for Solid-State Illumination and Technologies) program started a project for an LED lamp reliability analysis a few years ago under the auspices of an industry consortium [31–33]. The aim of the ASSIST project was to create a test method that could be used to predict LED lamp life in any application by measuring on/off power cycling with enough dwell time and taking into account both catastrophic and parametric failure forms.

The results of the study revealed that LED lamps have both catastrophic and parametric failure behaviors. Maximum operating temperature affects the lumen depreciation rate and parametric failure time, and on/off cycling promotes catastrophic failure. The study found that on/off switching shortens the life of LED lamps; most of the lamps failed catastrophically before the LED light output reached the 70% mark, or L70 standard, in comparison to the initial value.

Manufacturers desire shorter test times to determine product lifetimes. According to ASSIST, the first step in implementing a test for LED lamps is to determine an optimal temperature for reliability testing and test time such that LED lamps fail due to failure modes that are common in the applications of interest. Overstressing can introduce a new failure mechanism that may not be present in traditional applications, causing LED lamp lifetimes to be underestimated.

ASSIST, for example, suggested a research protocol in which the time-averaged temperature is between 75 and 85°C and the total testing time is less than 3000 h. This suggestion would need to be confirmed by testing multiple LED lamps on the market. When sufficient data (time to fails with respect to the setup parameters) are collected, the predictive model or projection can be established. The findings of this study showed that by knowing the LED temperature and switching cycle, a faster test time procedure can be established to accurately predict LED lamp in any application.

All these reliability standards development activities demonstrate the importance of LED lighting reliability measurement. They also reflect the challenges and difficulty involved in standard development. Basically, writing a standard is not the goal, but how the standard can be effectively used and assisted LED lighting commercialization is the purpose. Therefore, once developed, reliability standards must be widely accepted and practiced by the industry.

### 9.3.3 JEDEC test standards

JEDEC (Joint Electron Device Engineering Council) standards, common standards used in the electronics and semiconductor industries for over 50 years [34,35], are also employed in LEDs reliability testing, especially for outdoor LED lighting applications such as street and area lighting, where the life requirements of solid-state lighting (SSL) luminaires exceed

50,000 h and LEDs are subjected to harsh environmental conditions. The LED packages within luminaires must pass a series of reliability or certification tests to assure that they can meet specified lumen performance over the stated life under a given environmental stress. Examples are the US Department of Energy-sponsored Commercial Building Energy Alliance (CBEA) Parking Structure Lighting Performance Specification and the DOE-sponsored Municipal Solid-State Street Lighting Consortium (MSSLC) Model Specification for LED Roadway Luminaires.

In automotive LED lighting applications, the need to test an LED package's reliability against environmental stress was first implemented. The Component Technical Committee of the Automotive Electronics Council (AEC) has produced a list of LED package tests. The AEC Q101 standard, titled "Stress Test Qualification for Discrete Semiconductors," contains a list of JEDEC standards that can be referred to when testing specifications are required.

Given that JEDEC provides a variety of standards for testing LEDs under various stresses, let us now discuss the specific JEDEC standards for coping with the various stresses that LEDs will face.

#### *9.3.3.1 Environmental stress*

Three types of LED stresses are used in JEDEC tests applicable to LED packages for outdoor lighting applications. Temperature, humidity, and salt corrosion fall into the first category of environmental stress. When in service, the LEDs are checked at extreme high and low temperatures, as well as in a storage situation (JESD 22A108-D, JESD 22 A101-C, and JESD 22 A108-C) [36–38]. The LEDs are also measured in high-temperature and high-humidity environments, as well as in conditions where the temperature is constantly cycled between high and low while the LEDs are turned off and on. These tests are used to determine how temperature and humidity affect the device's mechanical, electrical, and optical performance.

A thermal shock test (JESD 22-A104) is also available [39]. This is a more rigorous temperature-related test in which LEDs are moved from extremely hot environments (as high as 85–150°C) to extremely cold environments (−40 to −65°C) in under 20 seconds. Typically, these tests are used to detect electrical link mechanical failures inside LED packages.

Another JEDEC environmental test (JESD22-A107C) [40] that can be used to assess corrosion caused by salt is the JESD22-A107C. Salt corrosion of LED components can occur in both automotive and outdoor lighting environments. Such test is needed because product quality and component performance in a corrosive environment can differ greatly from one manufacturer to another. Corrosion can cause electrical connections to fail mechanically and degrade optically. In practice, LED manufacturers have discovered that corrosion can be caused by other chemicals in the atmosphere or by the outgassing of other components within the SSL luminaire. This may cause the LEDs to fail or cause them to perform poorly. Since the technologies and materials used in LED packages and luminaires differ, LED or SSL manufacturers typically decide the tests for other forms of chemical corrosion.

#### 9.3.3.2 Mechanical stress

Mechanical stress, the second form of LED-package stress, includes mechanical shock (JESD 22 B104-C) [41] and variable-frequency vibration (JESD 22-B103) [42]. These tests are needed because luminaires in automotive or highway lighting applications will be exposed to mechanical shock and vibrations during installation and use. Many mechanical interfaces and connections in an LED package must be able to withstand repetitive and extreme impact or vibrations over time. The mechanical behavior or defects of LED packages are identified using these stress tests.

#### 9.3.3.3 Stress due to assembly process

The third group looks at the stresses that occur during the LED assembly process. Depending on the LED's design, such as surface-mounted versus through-hole-mounted, this may require two or three tests. The tests are used to determine the sensitivity of LEDs to moisture or reflow, especially their resistant to soldering heat. In most cases, LED packages are soldered to a board and then inserted into the final SSL product assembly. LEDs may be exposed to extreme temperatures during the LED-to-board assembly phase. These tests are the foundation of LED package requirements, and solder temperature profiles are often used in datasheets for LEDs. It's clear that LEDs must withstand the initial stress of assembly when they're placed on the boards.

#### 9.3.3.4 Additional JEDEC tests

Additional JEDEC specifications, such as an electrostatic discharge (ESD) test, are used in the LED stress tests. One test uses the human body model (HBM) to determine ESD sensitivity, determining an LED package's susceptibility to harm or deterioration due to human handling.

The JEDEC tests are commonly used as standard certification tests prior to product launch by reputable LED manufacturers, first in LED automotive lighting applications and now for outdoor, street, and area lighting applications. LEDs that have gone through and passed these JEDEC tests have been shown to be more reliable in terms of lumen production maintenance and other efficiency metrics.

As can be seen from the previous discussions, there are numerous organizations working toward the reliability test standards for LEDs testing and lifetime estimation, few of the standards are widely used. These standards are as follows:

- *ANSI Standard*: ANSI/IES RP-16-10: Its aim is to define solid-state lighting devices and their components so that everyone understands what they're talking about [43].
- *ANSI Standard*: ANSI C78.377-2011: It is formulated to communicate the indoor lighting SSL products color characteristics to SSL producers and users [44,45].
- *IES LM-79-08*: It provides procedures for reproducible measurements of photometry, color and electrical characteristics of SSL products [46,47].
- *IES LM-80-08*: It provides methods of the measurement of lumen maintenance of LED packages, arrays and modules [28,48,49].
- *IES TM-21-1*: It provides a calculation tool to interpret the data collected from LM-80 testing; to provide users with lumen maintenance life (e.g., L70) projection, or to predict estimated lumen output values at a given time duration; to interpolate lumen maintenance behaviors for the in situ temperature (different from testing temperature) [50,51].
- *IES LM-82-11*: It is to describe the procedures in performing reproducible measurements of LED light engines and integrated LED lamps, at any given temperature for the performance characteristics (total luminous flux, electrical power, etc.) [52,53].
- *UL8750*: It is to provide safety requirements for LED equipment that is an integral part of a luminaire or other equipment and which operates in the visible light spectrum between 400 and 700 nm. The requirements also cover the component parts of the LED equipment, including LED drivers, controllers, arrays, modules, and packages as defined within this standard [54,55].
- *IES LM-84*: It provides the method for measurement of lumen and color maintenance of LED lamps, light engines, and LED luminaires. To describe the procedures to obtain uniform and reproducible lumen and

color maintenance measurements under standard operating conditions [56,57].

- *IES LM-85*: It addresses the measurements for high-power LEDs that require heat sink for their normal operation and include white LEDs as well as single color LEDs [58,59].

More information on the abovementioned standards and their testing procedures can be found in Refs. [28,43–59].

## 9.4 Challenges in extrapolation of LED reliability test data to current operating conditions

Extrapolation of reliability test results to standard operating conditions is important since most reliability tests are performed at higher stress levels. Peck's model, Eyring model, Coffin–Manson model, and Arrhenius model are some of the acceleration models commonly used for extrapolation of electronic products under stress [60–63]. However, since LED degradation mechanisms evolved with time and they also undergo lumen degradation recovery, as discussed in the previous chapters, some of these models may not be applicable to predict LED lifetime, in contrast to other electronic devices.

There is currently no good model for extrapolating LED lifetime for outdoor applications, and a new method for extrapolating LED reliability data to normal operating conditions, including humidity, is required. The new TM-21 and LM80 are solely concerned with LED thermal degradation.

## 9.5 Need for additional reliability test standards for extended applications of LEDs

Two problems remain after a decade of developing reliability standards. The first question is how the industry will standardize SSL outdoor lighting specifications in LED package-level certification or reliability tests. Although some JEDEC tests are listed or referenced in the CBEA and MSSLC specifications, the precise criteria are still not defined. For example, failure criteria are not clearly defined, and test conditions and durations are not standardized. Despite some debate during the creation of the above requirements, the final response remains a mystery.

The second, more difficult question to answer is how to quantitatively compare the LED package JEDEC test results to their predicted reliability in terms of rated life in hours. If the LED packages passed all of the above JEDEC measures, which are usually calculated in 1000 h or comparable time periods under extreme stress, the question of how many hours of real-world service can be claimed for the LED's rated life remains.

In many SSL outdoor lighting applications, the rated life of LEDs is usually specified as 50,000 h or longer. There could be a correlation between this rated life and 1000 h of extreme stress-test performance, and this could be a statistical indicator. However, no standard method for detecting the connection has yet been established. The industry needs to develop a standard protocol based on the practice of testing LEDs according to different criteria and projecting test results to LED package rated life.

## 9.6 Summary

This chapter dealt in depth with the various LED reliability standards and the organizations working toward the development of these standards. IEC, IES, ASSIST, JEDEC, etc., have provided various reliability tests standards for LED reliability under various stress conditions. However, due to the complex structure of LEDs and ever-improving materials and structural design, there is a need for modification in the present reliability standards. LED's overall lumen degradation is generally dependent on multiple degradation mechanisms, and there is also a need for reliability tests, which can integrate the dependency of failures of the various parts of the luminary on the overall lumen degradation. This chapter also showed that present extrapolation techniques are not applicable to obtain normal condition LED lifetime from a testing data as the degradation mechanism varies for LEDs under different stress conditions and the failure mechanisms also evolve over time.

## References


[1] M. Wei, K.W. Houser, Status of solid-state lighting based on entries to the 2010 US DOE Next Generation Luminaire competition, Leukos 8 (4) (2012) 237–259.

[2] A. Wilkerson, G.P. Sullivan, R.G. Davis, S. Safranek, Yuma Border Patrol Lighting Retrofit: Final LED System Performance Assessment of Trial and Full Installation (No. PNNL-27293), Pacific Northwest National Lab (PNNL), Richland, WA (United States), 2018.

[3] E.E. Richman, The Elusive "Life" of LEDs: How TM-21 Contributes to the Solution (No. PNNL-SA-83710), Pacific Northwest National Lab (PNNL), Richland, WA (United States), 2011.

[4] H. Ganev, T.Q. Khanh, Degradation of high-power LEDs—luminous flux and TM 21 life time, in: 2014 11th China International Forum on Solid State Lighting (SSLCHINA), IEEE, November 2014, pp. 141–143.

[5] E. Richman, Understanding LED tests: IES LM-79, LM-80, and TM-21, in: DOE SSL Workshop, July 2011.
[6] R.A. Guisso, M.F. Righi, E.M. Sá, R.A. Pinto, V.C. Bender, T.B. Marchesan, An extended design methodology for LED lighting systems including lifetime estimation, IEEE Transactions on Electron Devices 63 (12) (2016) 4852–4859.
[7] S.S. Letha, M. Bollen, LED Life Assessment by Accelerated Test Methods-A Review, 2021.
[8] A. Jayawardena, N. Narendran, Analysis of electrical parameters of InGaN-based LED packages with aging, Microelectronics Reliability 66 (2016) 22–31.
[9] J.R. Tuenge, M.E. Poplawski, PoE Lighting System Energy Reporting Study, Part 1 (No. PNNL-26284), Pacific Northwest National Lab (PNNL), Richland, WA (United States), 2017.
[10] M.H. Chang, D. Das, P.V. Varde, M. Pecht, Light emitting diodes reliability review, Microelectronics Reliability 52 (5) (2012) 762–782.
[11] D. Mellinger, P. LC, Alliance to save energy (funded by GE current, a Daintree company), in Partnership with the Design Lights Consortium (DLC).
[12] A. Hinge, S.E. Partnerships, S. Dagher, N.E.E. Partnerships, An Approach to Promoting Smart Lighting for Industry.
[13] M. Royer, R. Tuttle, S. Rosenfeld, N. Miller, Color Maintenance of LEDs in Laboratory and Field Applications, 2013.
[14] A. Schwarzenegger, Lighting California's Future: Advanced Energy-Efficient LED Lighting for Residential and Commercial Applications, California Energy Commission, 2010.
[15] R.F. Service, The Quest for White LEDs Hits the Home Stretch, 2009.
[16] C.M. Tan, P. Singh, Time evolution degradation physics in high power white LEDs under high temperature-humidity conditions, IEEE Transactions on Device and Materials Reliability 14 (2) (2014) 742–750.
[17] C. MingTan, P. Singh, A review on degradation physics of high power LEDs in outdoor applications, in: 2016 IEEE 23rd International Symposium on the Physical and Failure Analysis of Integrated Circuits (IPFA), IEEE, July 2016, pp. 383–386.
[18] M.G. Pecht, M.H. Chang, Failure Mechanisms and Reliability Issues in LEDs. Solid State Lighting Reliability, 2013, pp. 43–110.
[19] J. Fan, K.C. Yung, M. Pecht, Physics-of-failure-based prognostics and health management for high-power white light-emitting diode lighting, IEEE Transactions on Device and Materials Reliability 11 (3) (2011) 407–416.
[20] E.E. Storage, International Electrotechnical Commission (IEC). Geneva, Switzerland, 2011.
[21] M. Voytchev, R. Radev, P. Chiaro, I. Thomson, C. Dray, J. Li, IEC International Standards under Development for Radiation-Generating Devices (No. UCRL-PROC-237155), Lawrence Livermore National Lab.(LLNL), Livermore, CA (United States), 2007.
[22] S.B. Dhia, A. Boyer, B. Li, A.C. Ndoye, Characterisation of electromagnetic compatibility drifts of nanoscale integrated circuit after accelerated life tests, Electronics Letters 46 (4) (2010) 278–280.
[23] C. Jung, T. Zahner, Reliability management of a light-emitting diode for automotive applications, in: Solid State Lighting Reliability Part 2, Springer, Cham, 2018, pp. 549–572.
[24] J. Huang, D.S. Golubović, S. Koh, D. Yang, X. Li, X. Fan, G.Q. Zhang, Degradation mechanisms of mid-power white-light LEDs under high-temperature–humidity conditions, IEEE Transactions on Device and Materials Reliability 15 (2) (2015) 220–228.
[25] C. Zhaohui, Z. Qin, W. Kai, L. Xiaobing, L. Sheng, Reliability test and failure analysis of high power LED packages, Journal of Semiconductors 32 (1) (2011) 014007.

[26] J. Li, J. Wang, X. Yi, Z. Liu, T. Wei, J. Yan, B. Xue, Reliability analysis of group III nitride LEDs devices, in: III-nitrides Light Emitting Diodes: Technology and Applications, Springer, Singapore, 2020, pp. 203–227.
[27] P. Lall, P. Sakalaukus, L. Davis, Reliability of solid-state lighting electrical drivers subjected to WHTOL accelerated aging, in: Fourteenth Intersociety Conference on Thermal and Thermomechanical Phenomena in Electronic Systems (ITherm), IEEE, May 2014, pp. 1164–1170.
[28] J. Hegedüs, G. Hantos, A. Poppe, Reliability issues of mid-power LEDs, in: 2019 25th International Workshop on Thermal Investigations of ICs and Systems (THERMINIC), IEEE, September 2019, pp. 1–6.
[29] W. Tang, J.G. Liu, C. Shen, Blue light hazard optimization for high quality white LEDs, IEEE Photonics Journal 10 (5) (2018) 1–10.
[30] B. Liang, Z. Wang, C. Qian, Y. Ren, B. Sun, D. Yang, J. Fan, Investigation of step-stress accelerated degradation test strategy for ultraviolet light emitting diodes, Materials 12 (19) (2019) 3119.
[31] J.D. Bullough, K. Sweater Hickcox, T.R. Klein, N. Narendran, Effects of flicker characteristics from solid-state lighting on detection, acceptability and comfort, Lighting Research & Technology 43 (3) (2011) 337–348.
[32] J.P. Freyssinier, M. Rea, A two-metric proposal to specify the color-rendering properties of light sources for retail lighting, in: Tenth International Conference on Solid State Lighting, vol 7784, International Society for Optics and Photonics, August 2010, p. 77840V.
[33] National Research Council, Assessment of Advanced Solid-State Lighting, National Academies Press, 2013.
[34] D. Edwards, Development of JEDEC standard thermal measurement test boards, IEEE Transactions on Components, Packaging, and Manufacturing Technology: A 19 (4) (1996) 478–485.
[35] R. Chen, M.H. Wang, S. Lee, J. Ho, I. Hu, S. Shih, New/Old JEDEC board level drop reliability test standards evaluation: measurement and simulation study, in: 2017 IEEE CPMT Symposium Japan (ICSJ), IEEE, November 2017, pp. 211–215.
[36] A. Poppe, Testing of Power LEDs: The Latest Thermal Testing Standards from JEDEC, 2013.
[37] G. Hantos, J. Hegedüs, K-factor calibration issues of high power LEDs, in: 2017 23rd International Workshop on Thermal Investigations of ICs and Systems (THERMINIC), IEEE, September 2017, pp. 1–6.
[38] B. Marovic, A. Francois-Saint-Cyr, Transient thermal measurement of electronic components and radiometric characterization of LEDs, in: Proceedings of the FISITA 2012 World Automotive Congress, Springer, Berlin, Heidelberg, 2013, pp. 1669–1676.
[39] G.Q. Lu, J.N. Calata, G. Lei, X. Chen, Low-temperature and pressureless sintering technology for high-performance and high-temperature interconnection of semiconductor devices, in: 2007 International Conference on Thermal, Mechanical and Multi-Physics Simulation Experiments in Microelectronics and Micro-systems. EuroSime 2007, IEEE, April 2007, pp. 1–5.
[40] M.N. Nor Ilyana, Evaluation of Corrosive Environment Impact on Low Cost Fe/Al ADDED SAC105 Solder alloys/Nor Ilyana Muhd Nordin (Doctoral Dissertation, University of Malaya), 2017.
[41] A. Farris, Drop Impact Reliability Testing Lead-Free Chip Scale Packages, 2008, p. 21. Master's Theses and Project Reports.
[42] S. Tuukkanen, T. Julin, V. Rantanen, M. Zakrzewski, P. Moilanen, K.E. Lilja, S. Rajala, Solution-processible electrode materials for a heat-sensitive piezoelectric thin-film sensor, Synthetic Metals 162 (21–22) (2012) 1987–1995.

[43] K.W. Houser, Light, light, light, LEUKOS 8 (1) (2011) 5–7.
[44] A. Baker, ANSI C78. 377-201x Revision: New Nominal CCTs Smaller Quadrangles and Circles, 2014.
[45] M.R. Luo, G. Cui, M. Georgoula, Colour difference evaluation for white light sources, Lighting Research & Technology 47 (3) (2015) 360–369.
[46] I. LM, Electrical and Photometric Measurements of Solid-State Lighting Products, 2008.
[47] G. Dutt, N. Herrick, S. Kumar, P. Vishwanath, R. Pandher, Die Attach Materials and LED Functional Performance.
[48] J. Fan, K.C. Yung, M. Pecht, Lifetime estimation of high-power white LED using degradation-data-driven method, IEEE Transactions on Device and Materials Reliability 12 (2) (2012) 470–477.
[49] H.B. Fan, X.P. Li, J.X. Shen, M. Chen, An effective prediction method for LED lumen maintenance, in: 2012 13th International Conference on Electronic Packaging Technology & High Density Packaging, IEEE, August 2012, pp. 1560–1563.
[50] L. Turunen, K. Mönkkönen, LedTrix-Led-teknologian Osaamispankki, Loppuraportti, 2014.
[51] L. Davis, K. Rountree, K. Mills, Luminous Flux and Chromaticity Maintenance for Select High-Power Color Light-Emitting Diodes (No. DOE/EE-1801), RTI International, Research Triangle Park, NC (United States), 2018.
[52] Z. Zhike, C. Suming, C. Shasha, S. Tingting, L. Yuanhui, Comparison of life testing standards for LED lighting products, in: 2019 16th China International Forum on Solid State Lighting & 2019 International Forum on Wide Bandgap Semiconductors China (SSLChina: IFWS), IEEE, November 2019, pp. 225–230.
[53] H.L. Ke, Q. Sun, J. Zhao, H.X. Zhang, L. Jing, Y. Wang, J. Hao, Junction temperature estimation for LED lamp with forward voltage method, in: 2016 17th International Conference on Thermal, Mechanical and Multi-Physics Simulation and Experiments in Microelectronics and Microsystems (EuroSimE), IEEE, April 2016, pp. 1–4.
[54] S.S. Bar, Z.L. Specifications, General.
[55] C. Ye, P. Das, S.K. Sahoo, Peak current control of multichannel LED driver with selective dimming, IEEE Transactions on Industrial Electronics 66 (5) (2018) 3446–3457.
[56] N. Narendran, Y.W. Liu, X. Mou, D.R. Thotagamuwa, O.V.M. Eshwarage, Projecting LED product life based on application, in: Fifteenth International Conference on Solid State Lighting and LED-Based Illumination Systems, vol 9954, International Society for Optics and Photonics, September 2016, p. 99540G.
[57] M. Royer, Lumen maintenance and light loss factors: consequences of current design practices for LEDs, Leukos 10 (2) (2014) 77–86.
[58] J. Jiao, IES LM-85 homogenizes high-power LED measurement and characterization, LEDs Magazine 11 (7) (2014) 55–58.
[59] C.C. Miller, A. Jillavenkatesa, Pathway to developing a UV-C standard—a guide to international standards development, IUVA News 20 (2018) 22–26.
[60] H. Caruso, A. Dasgupta, A fundamental overview of accelerated testing analytical models, Journal of the IEST 41 (1) (1998) 16–20.
[61] L.A. Escobar, W.Q. Meeker, A review of accelerated test models, Statistical Science (2006) 552–577.
[62] P. Viswanadham, P. Singh, Failure models, in: Failure Modes and Mechanisms in Electronic Packages, Springer, Boston, MA, 1998, pp. 283–309.
[63] Z. Illyefalvi-Vitez, P. Nemeth, P. Bojta, Failure and acceleration models for MCM-Ls tested by HAST, in: 52nd Electronic Components and Technology Conference 2002. (Cat. No. 02CH37345), IEEE, May 2002, pp. 480–483.

# Index

*Note*: Page numbers followed by "*f*" indicate figures and "*t*" indicate tables.

Kolibri 360 GmbH
Ferdinand-Jühlke-Straße 7
99095 Erfurt, Deutschland
produktsicherheit@kolibri360.de